PLASMA HARMONICS

PLASMA HARMONICS

Rashid A. Ganeev

Pan Stanford Publishing

Published by

Pan Stanford Publishing Pte. Ltd.
Penthouse Level, Suntec Tower 3
8 Temasek Boulevard
Singapore 038988

Email: editorial@panstanford.com
Web: www.panstanford.com

British Library Cataloguing-in-Publication Data
A catalogue record for this book is available from the British Library.

Plasma Harmonics

ISBN 978-981-4463-30-0 (Hardcover)
ISBN 978-981-4463-31-7 (eBook)

Printed in the USA

To my parents, wife, son, and daughter

. . . А с этою плазмой
Дойдешь до маразма.
И это довольно почетно!

—В. С. Высоцкий

. . . And this plasma
Makes you crazy.
However, it is a privilege!

—V. S. Vysotsky

Contents

Preface

I wrote this book two years after completing the previous monograph on the high-order harmonic generation of laser radiation in plasma plumes. The reasons for writing another book on the same topic were the incompleteness of data presented in the previous publication, appearance of new results on plasma harmonics generation, and willingness to provide a thorough analysis of some nonlinear optical processes that occur in laser plasma during the propagation of ultrashort laser pulses through this medium. Although I have tried to not repeat my previous findings in this book, some basic thoughts, which remain the same for both books, have been presented and emphasized. As I have already stated, the objective of writing this book is to present the missing parts of this field of nonlinear optics that have motivated me to show new theoretical considerations of plasma harmonics, advanced methods of plasma formation, and recent experimental findings of various new schemes of the *high-order harmonic generation of laser radiation in laser-produced plasma plumes*, which I have simply dubbed "plasma harmonics" in this book. Some incorrectness in this term is probably compensated by the shortness and attractiveness of the book title. Indeed, since the first experiments of high-order harmonic generation in gaseous media (gas jets and gas cells), the term "gas harmonics" has often appeared in scientific literature along with the term "surface harmonics," which refers to another process of high-order harmonic generation during specular reflection of strong laser fields from the surfaces of various materials. Since most studies on harmonics generated in specially prepared plasma plumes during last few years were performed with my participation (both experimental research in Japan, India, Canada, United Kingdom, Germany, Spain, Uzbekistan, etc., and theoretical studies in Bosnia and Herzegovina, Germany, Uzbekistan, Russia, etc.), I feel the privilege to dub this field of nonlinear spectroscopy as plasma harmonics.

So, what distinguishes this book from my previous publication, *High-Order Harmonic Generation in Laser Plasma Plumes* (Imperial College Press, London, 2012)? First, I discuss why at all plasma harmonics may be of interest to us and in which areas of physics it can be applied. I introduce both old and new theories related with this field and give a brief historic overview on how the interest in this field of nonlinear optics dropped after the first attempts of high-order harmonic generation from plasma plumes in the first half of 1990s, but revived in 2005 when new approaches in the technique of coherent extreme ultraviolet radiation generation considerably pushed ahead this area of study. I show both disadvantages and advantages of harmonic generation in plasma, which allowed to make this approach concurrent to the commonly accepted gas harmonic technique. The book also presents an analysis of the low- and high-order plasma harmonics generated under different experimental conditions and using various (nanosecond, picosecond, and femtosecond) lasers and then discusses the new approaches in plasma harmonic studies. An important feature of these studies is the characterization of plasma plumes and plasma harmonics, which was not fully described in my previous book. I also present the recent achievements in plasma harmonics and discuss the perspectives of this interesting field of nonlinear spectroscopy of matter.

Overall, these studies turned from being interested in the creation of new efficient sources of coherent extreme ultraviolet sources to become a tool for studying the nonlinear optical, orientational, and morphological properties of various materials that could be ablated from solid surfaces. Currently, there is worldwide interest in plasma harmonics, since the application of plasma plumes for the analysis of various properties of matter using the methods of high-order nonlinear spectroscopy—analogous to the gas harmonics study of some orientational features of a few molecular gases—can answer many questions about the structural properties of materials. These studies are already under way in various laboratories worldwide. It is expected that plasma harmonic generation, which allows the application of thousands of solids, contrary to few tens of gaseous media used in gas harmonic approach, may further push this field of knowledge toward the better understanding of the features of various materials.

My visits to several leading laboratories working on plasma harmonics led to fruitful communication with PhD students, postdocs, and many colleagues involved in these studies. I gratefully acknowledge their collaboration, because while familiarizing them with the basics of this technique, I, too, learned many new methods and became acquainted with the new and advanced equipment and technologies. I would like to thank H. Kuroda, M. Suzuki, M. Baba, F. Mitani (Saitama Medical University, Japan); T. Ozaki, L. B. Elouga Bom (Institut National de la Recherche Scientifique, Canada); P. D. Gupta, P. A. Naik, H. Singhal, J. A. Chakera, M. Kumar, U. Chakravarty, M. Tayyab, R. A. Khan, M. Raghuramaiah (Raja Ramanna Centre for Advanced Technology, India); J. P. Marangos, J. W. G. Tisch, C. Hutchison, T. Witting, F. Frank, A. Zaïr, Z. Abdelrahman, F. McGrath (Imperial College, United Kingdom); M. Castillejo, M. Oujja, M. Sanz, I. López-Quintás, M. Martin (Instituto de Química Física Rocasolano, Spain); H. Zacharias, J. Zheng, M. Wöstmann, H. Witte (Westfälische Wilhelms-Universität, Germany); T. Usmanov, I. A. Kulagin, V. I. Redkorechev, V. V. Gorbushin, R. I. Tugushev, G. S. Boltaev, N. K. Satlikov (Institute of Ion-Plasma and Laser Technologies, Uzbekistan); M. Danailov (ELETTRA, Italy); B. A. Zon, N. L. Manakov, M. V. Frolov (Voronezh State University, Russia); D. B. Milošević (University of Sarajevo, Bosnia and Herzegovina); M. Lein, M. Tudorovskaya (Leibniz Universität Hannover, Germany); E. Fiordilino (University of Palermo, Italy); V. V. Strelkov, M. A. Khokhlova (General Physics Institute, Russia); M. K. Kodirov, P. V. Redkin (Samarkand State University, Uzbekistan); A. V. Andreev, S. Y. Stremoukhov, and O. A. Shoutova (Moscow State University, Russia) for their support in the development of plasma harmonics technique. A tracking of the list of people and institutions involved in this research just shows the interest in this relatively new field of optics. It is important to note that, recently, several laboratories in the United States, France, Spain, Korea, Ireland, Italy, and Japan have become involved in plasma harmonics studies.

Probably, my wife, Lidiya, son, Timur, and daughter, Dina, have suffered when I could not take them along with me on my long trips abroad. Their patience and love always followed my love for them, which helped me overcome the various obstacles of the life of a scientific traveler. I thank them for that.

Rashid A. Ganeev
Saitama, Japan
Spring 2014

Chapter 1

Why Plasma Harmonics?

1.1 A Very Brief Introduction

Indeed, why plasma harmonics? Why we need plasma harmonics while we have a well-elaborated technique of frequency conversion of laser radiation through high-order harmonic generation (HHG) in gaseous media? One can remind that there is also another technique of HHG, surface harmonic generation. Both of them have a relatively long history of development, improvement, and application for various needs, and it seemed that those two methods fully satisfied scientific community from the point of view of creation of the sources of coherent extreme ultraviolet (XUV) radiation.

From the common point of view, the developments in our life should follow in accordance to simplicity and economy. Ptolemy stated [1], "We consider it a good principle to explain the phenomena by the simplest hypothesis possible." This philosophical vision of everyday life later was developed in the principle of Ockham's razor as a principle of parsimony, economy, or succinctness used in logic and problem-solving. It states that among competing hypotheses, the one that makes the fewest assumptions should be selected. Particularly, Ockham cited the principle of economy, *Frustra fit per plura quod potest fieri per pauciora* [It is futile to do with more

Plasma Harmonics
Rashid A. Ganeev

ISBN 978-981-4463-30-0 (Hardcover), 978-981-4463-31-7 (eBook)
www.panstanford.com

things that which can be done with fewer]. It works well in life, but not in science. Moreover, the science itself is an example of broken Ockham's razor principle, as new, more sophisticated ideas, which contradicted with commonly accepted and simplest explanations and approaches in description of the basics of our universe, finally completely changed our vision of the surrounding world.

But let us return from the heights of philosophical heaven to our Earth and start considering the new approach in optics, which have no aim to replace existing methods of the development of coherent short-wavelength sources, but just feels good enough to find its own place in this field of human knowledge. Is the principle "It is futile to do with more things that which can be done with fewer" applicable in science, and, particularly, in nonlinear optics? To answer on the question put in the title of this chapter, let us first briefly remind the history of gas and surface HHG, as well as best achievements of these techniques.

I would like to recall that the earliest observations of HHG in gases were carried out in the second half of eighties using picosecond Nd:YAG lasers [2, 3] (as well as an excimer laser at 248 nm [4]). The harmonics from different gases up to the 21st and 33rd orders of 1064 nm radiation were reported at an intensity of 3×10^{13} W/cm^2, which led to an enormous growth of interest in this area of nonlinear optics. Those studies have demonstrated that application of gases as nonlinear media can be used as an advanced method for generation of coherent XUV radiation using picosecond-driving pulses. Those early developments were further transformed in the field of gas HHG spectroscopy when new, predominantly, femtosecond lasers became involved in this field of studies. Currently, the harmonics up to the 5000th orders were reported, although most of recent studies are related with the development of attosecond sources of laser radiation through gas HHG, analysis of orientational features of some gaseous molecules through the study of variable harmonic spectra from these species, as well as applications of gas harmonics for surface science, biology, medicine, and different branches of physics and chemistry. The attractiveness of this method is based on availability of moderate level femtosecond lasers in many laboratories worldwide and simplicity of handling the gas jet technique.

Another method, surface HHG, which is based on cardinally different physical principles than gas HHG, is less popular due to sophisticated equipment required for its implementation, however, well elaborated and used in advanced laboratories. Very high fluences and intensities (of the order of 10^{18} W/cm^2 or higher) and, most important, very high contrast ratios between driving pulse and prepulse already existing in any laser system are the main requirements for this technique. Not much laboratories can afford these conditions for surface HHG. Nevertheless, high cutoffs (of the order of 2000th harmonics) and high conversion efficiencies for lowest orders of harmonics were reported using this technique, though practical applications of this method are doubtful.

Other approaches in coherent XUV radiation generation include X-ray lasers [5–7] and free-electron lasers [8, 9], which are based on principles other than HHG of laser radiation by different means described above [10, 11]. Although the Web of Science search database gives approximately equal number of publications on those topics, the latter approach seems very attractive from the point of view of availability of the moderate energy/intensity lasers in many laboratories worldwide and less expenditure for the everyday use compared with two former methods. Other disadvantages of X-ray lasers are their poor spatial coherence and radiation divergence. With regard to free-electron lasers that generate radiation in the short-wavelength spectral range, there are only a limited number of sources available so far. Furthermore, the application of these lasers is largely limited by their high cost.

Returning back to HHG approaches, one has to note that there are many good publications showing a whole picture of developments of the gas HHG (e.g., the review [12]). The mechanisms of odd and even harmonics generation in the reflection of laser radiation from the surfaces are also frequently discussed in literature (see monograph [13] and review [14] and references therein). These HHG techniques are beyond the scopes of this book. My aim is to familiarize the reader with most recent approaches of harmonic generation in the XUV range with the use of an isotropic medium (plasma plume), which is different from that used previously in gas-jet sources and special gas-filled cells. Most recent monograph primarily discussing the experimental results of plasma harmonics

[15] has been published in 2012 and comprises many findings in this field. I will keep the terminology of HHG in laser-produced plasma plumes as "plasma harmonics," in accordance with explanation given in the Preface to this book.

So again, why plasma harmonics, when gas harmonics is a very elaborated technique, which, from one hand, allows generation of coherent XUV pulses and, from another hand, gives the opportunity in analysis of the orientational features of some molecular gases and provides the chances in generation and study of attosecond pulses generation and application? This question, without doubts, arisen in the first half of nineties of last century, when first attempts in generation of harmonics from the plasma plumes created during laser ablation of the surfaces of various metals were reported. The frustration caused by very unimpressive results shown in those first studies was the reason to ask themselves on the necessity in these experiments. In the following section of this chapter, I will discuss those results and reasons of why the proposed technique was forgotten for almost 10 years, and there were no laboratories dealing with the method during this period. This give up in plasma harmonics was followed with impressive achievements in creation of the sources of coherent XUV radiation through generation of gas harmonics and first reports of most impressive finding of that time–attosecond timescale of the pulses of harmonics. Probably, the Ockham's principle was a sufficient reason for this, as no ideas were disseminated among the laser community on how to overcome the problems of notoriously low conversion efficiency of this process, as well as low harmonic cutoffs (not exceeding twenties orders) comparing with excellent achievements in gas HHG.

It seems too early to talk about what the novelties were proposed during my first experiments with plasma plumes in 2005. Let us follow the logic of description of the historical developments of this field. In the next section, I will discuss some advantages, and, mostly, disadvantages, which were met by first investigators of plasma harmonics. I also was among them, and one can imagine my frustration in those results, which were far behind the best results of gas harmonics. So, let us add to this book a little bit of the history of plasma harmonic studies.

1.2 Early Stage of Plasma Harmonic Studies: Hopes and Frustrations

First studies of plasma HHG were aimed to extend the harmonic cutoff by using the particles with high ionization potentials. From this point of view, single-charged alkali metal ions (Li^+, Na^+, and K^+) should be a good choice, because they are isoelectric to neutral rare gases and they have higher ionization potentials than those of rare gas atoms. Akiyama et al. [16] used sub-picosecond KrF laser (0.5 ps, 6 mJ, 248 nm) for harmonic generation in the plasmas containing alkaline metal ions produced by 20 ns, 200 mJ, 248 nm radiation (Fig. 1.1a). The highest order of harmonics observed was the 13th in K^+, while the harmonics up to the 11th in Na^+, the ninth in Li^+ and He, and the seventh in Ar and Ne were also found. Among the nonlinear media used, the K^+ provided the highest intensity for all orders of harmonics (Fig. 1.1b).

The same team have used various other ions in laser-produced plasmas for the HHG in order to minimize the influence of ionization processes [17]. The highest harmonic order was the 21st harmonic in lead ions (Fig. 1.2a). The highest observed harmonic orders in various ions and neutral gases were found to be proportional to the ionization potentials of these species. Those experiments were carried out using extremely high intensity of KrF laser (0.5 ps, 3×10^{17} W/cm^2), while intensity of 20 ns heating pulse was in the range of 1×10^9 W/cm^2. Six solid targets were chosen with various atomic numbers other than three alkali metals as discussed above, namely boron ($Z = 5$), carbon (6), titanium (22), copper (29), tungsten (74), and lead (82). Figure 1.2b compiles the harmonic distributions using these six elements. For each species, a dominant charge state was evaluated using a collisional-radiative model.

Wahlström et al. [18] also studied HHG in laser-produced Na^+ and K^+ ions using 794 nm, 150 fs driving pulse and 100 ps, 1064 nm heating pulses (Fig. 1.3). The highest observed orders were not as high as expected from estimates based on the saturation intensities for these ions and on the focused intensities in the absence of defocusing. The spatial far-field distribution of the harmonic radiation was found to exhibit ring structures, which varied with

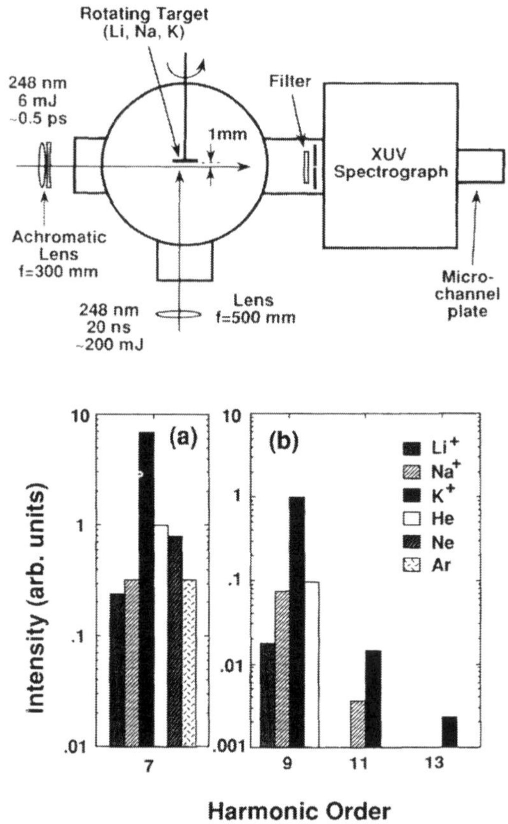

(a)

(b)

Figure 1.1 (a) Schematic diagram of the experimental setup. (b) Relative intensities of harmonics in the rare gases and alkali-metal ions. Reproduced from Ref. 16 with permission from American Physical Society. Copyright 1992.

the focusing conditions. They assumed that the limited extents of a plateau may be a consequence of ionization-induced defocusing of the high-power laser beam, reducing the peak intensity obtained in the medium. This defocusing is expected to be reduced by using short-wavelength laser for harmonic generation or by using a laser in the UV for the plasma formation, reducing a density of neutral atoms in the plasma.

Propagation effects in the HHG (up to the 13th order) of short-pulse KrF laser radiation in carbon vapor and low charged carbon

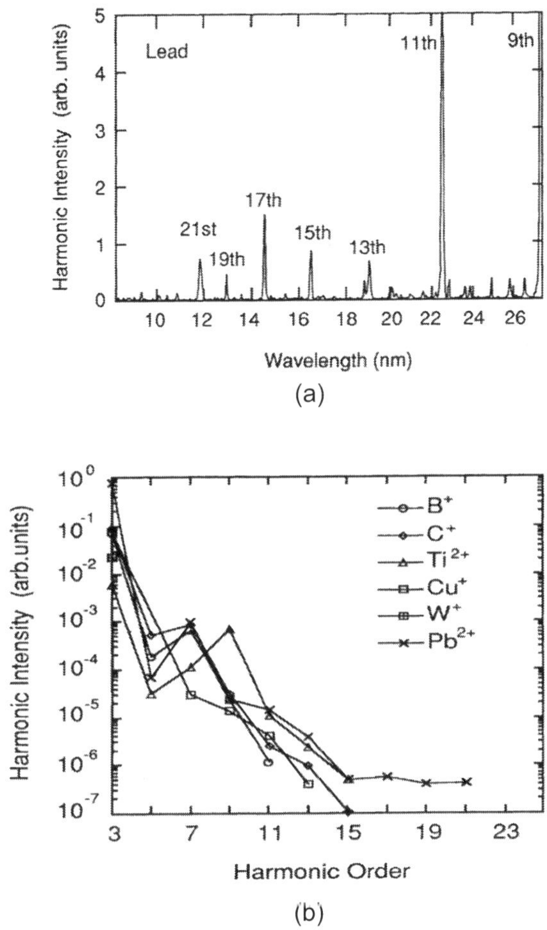

Figure 1.2 (a) Time-integrated high-order harmonic spectrum observed in a lead plasma. The spectral peaks are labeled by the harmonic orders. (b) High-order harmonic intensity distributions of six ionic species. Reproduced from Ref. 17 with permission from American Physical Society. Copyright 1993.

plasma have been studied by Theobald et al. [19]. They found that under used experimental conditions, high-order harmonics are generated mainly from neutral atoms. They also assumed that in a plasma, the sum-frequency generation of high-order harmonics was suppressed by an unfavorable positive phase-mismatch. In this

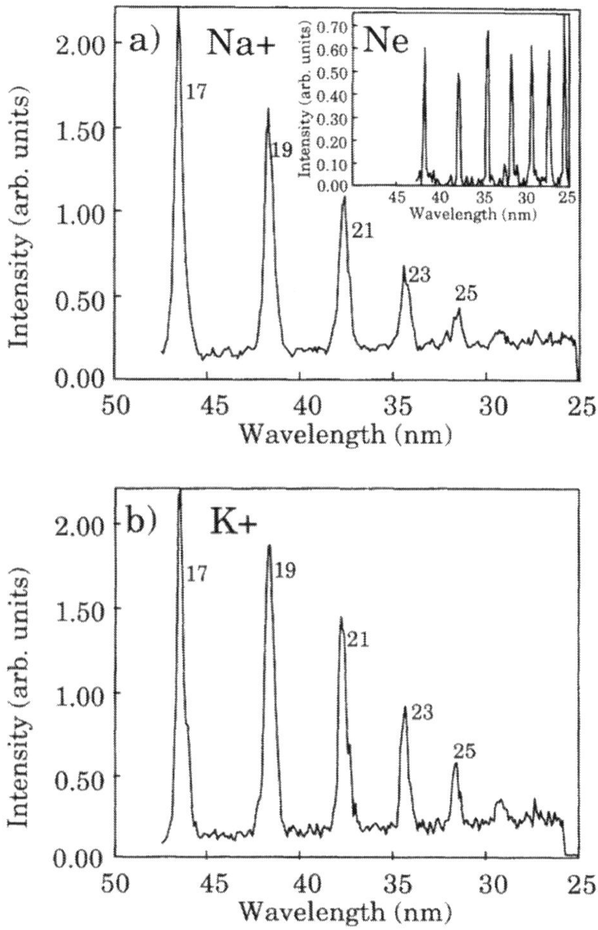

Figure 1.3 Harmonic spectra obtained (a) in Na^+ and (b) K^+. The inset in (a) is a part of spectrum obtained under similar conditions, but with the laser-produced plasma replaced with a pulsed nozzle producing a jet of neutral Ne atoms. Reproduced from Ref. 18 with permission from American Physical Society. Copyright 1995.

work, noncollinear phase-matched HHG by difference-frequency mixing in plasmas was also discussed. They suggested that the observed anomalous continuous growth of the fifth harmonic intensity with the plasma length originates from this noncollinear phase-matched difference-frequency mixing and defocusing of the laser radiation.

The possibility of optimization of high harmonics generation in a laser plasma when only one laser is used as a heating and driving source has been shown in [20]. An analysis of two different schemes of plasma excitation both in the atmosphere and in a vacuum chamber showed that the double-beam scheme has some advantages because of possibility of optimization of the delay conditions and excitation level. It was demonstrated that the train of picosecond pulses can serve as an effective source of production of the plasma for frequency conversion. Picosecond laser-produced plasma serves as an effective medium with the same nonlinear parameters as the one excited by nanosecond pulses used in previous investigations of harmonic generation in the plasma. The duration of the pulse train of picosecond Nd:glass laser allowed obtaining a corresponding delay between the heating and driving pulses when an optimal conversion to the short-wavelength radiation is realized.

In those experiments, odd harmonics generation (up to 11th order, $\lambda = 96$ nm) of IR laser radiation in a low-temperature atmospheric plasma and a plasma produced at the targets placed in a vacuum chamber were observed. Maximum efficiency (10^{-3}) was obtained for lowest order (third harmonic) nonlinear processes, while the efficiency of 11th harmonic was nearly 10^{-8}.

Krushelnick et al. [21] observed HHG from laser interactions with preformed underdense plasmas with subpicosecond ultraviolet laser pulses focused to intensities up to 10^{18} W/cm^2. The generation of seventh and ninth harmonics from aluminum plasmas was measured as well as harmonics to the 19th order from lithium–fluorine plasmas. The harmonic generation efficiency measured in these experiments was less than that from previous experiments that used neutral gases.

These experiments have shown that HHG is possible with intense laser pulses through the use of an exploding-foil preformed plasma and a multiplexed pulse train. In their experiments, harmonic generation was probably limited by the presence of free electrons in the nonlinear medium, which causes dispersion and phase mismatch (i.e., a significant level of ionization reduces the measured harmonic emission rates by more than an order of magnitude).

Nonetheless, it seemed that harmonic generation from ions is observable in plasmas if the polarizability of the ion in question is sufficiently large and the electron density is not too high. This was the first measurement of harmonic generation from highly ionized underdense aluminum plasmas, and the emission of seventh and ninth harmonics from these plasmas was recorded at laser intensities of 5×10^{17} W/cm^2, an intensity at which the efficiency of harmonic emission from initially neutral helium gas targets was similar. At low intensities, however, harmonic emission from neutral-gas targets was still significantly stronger than that from preformed plasma targets. The observation of seventh and ninth harmonics in underdense aluminum plasmas was probably due to a nonlinear polarization of the plasma ions. In fact, as the high-intensity laser pulse propagates through the plasma, the process of tunneling ionization may act to inhibit the production of harmonics if the generation mechanism can be regarded as a three-step process. The effect of ionization may also cause refractive defocusing, which prevents the laser beam from achieving high intensities in a plasma. This process can also create nonuniformities in the plasma that could then contribute to the phase mismatch. However, the use of a train of intense laser pulses may reduce the effects of such ionization-induced refraction on the trailing laser pulses because of the preionization effect of the first high-intensity laser pulse. At greater intensities, harmonic generation from high stages of ionization may, in fact, be advantageous because interactions with such ions are less susceptible to ionization-induced defocusing effects.

They suggested that, because an important potential application for this process is the efficient production of coherent XUV radiation, it may be useful to attempt to compensate for the large phase mismatches that occur in an ionized media through the use of atomic resonances or difference frequency mixing, or by using lasers of extremely high intensity so that the phenomenon of ionization suppression limits plasma production.

Those first experiments on HHG in the passage of laser radiation through the plasma produced during laser ablation of a solid

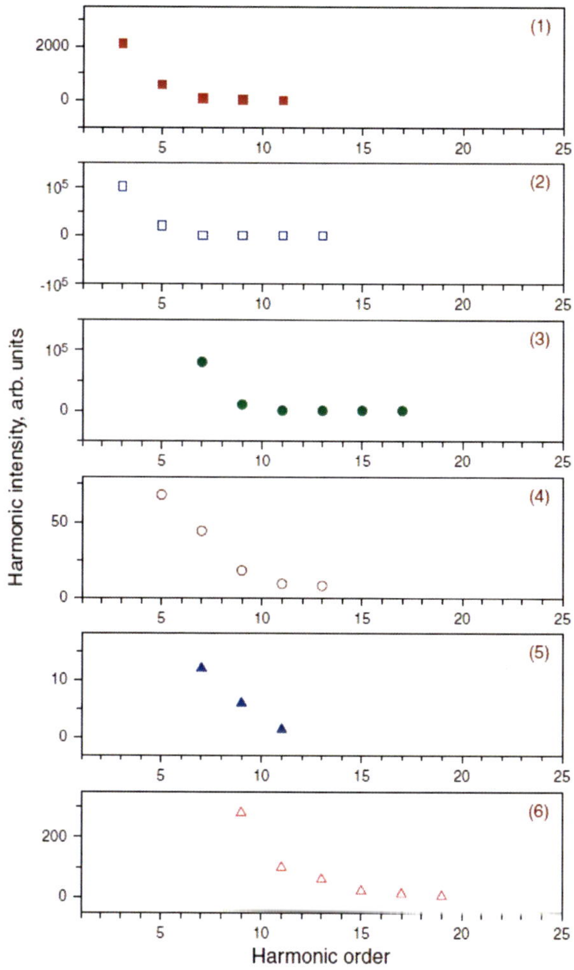

Figure 1.4 High-order harmonic spectra obtained from various laser-produced plasmas at the early stages of research (1992–1997) [22]. 1-Al, 2-Li, 3-C, 4-C, 5-Na, and 6-LiF.

target turned out to be much less successful. Above-presented data obtained with the use of highly excited plasmas containing multiply charged ions revealed several limiting factors, which did not permit generating harmonics of reasonable efficiency and sufficiently high orders. Those investigations have stopped at the demonstration of relatively low-order harmonics (see Fig. 1.4 that summarizes the best results attained during early plasma HHG studies [22]). These disadvantages, as well as the low conversion efficiency, led to the erosion of interest in this HHG technique, especially in comparison with the achievements involving gas HHG sources. From 1997 till 2005, there were no publications on plasma harmonics, as nobody believed in success in this field. All efforts of those groups, who initially tried to achieve some interesting results using this technique, were turned toward the gas HHG. The field of plasma harmonics was pronounced dead.

1.3 New Developments in Plasma Harmonics Studies: First Successes

I will tell you a story about my first observation of the high-order harmonics generating in laser plasma. This story may give some lessons to people, who are involved in science. Some occasional findings could meet us while we are dealing with other research. The luck and good fortune are important components of not only any adventurer but also any scientist. The goal of the latter is not to allow the new phenomenon appearing in his studies to be lost. If you are thinking about the new features of matter (in particular, high-order nonlinear optical properties of plasma plumes), or you have previously been involved in those studies, then it is natural to try to get something new while dealing with other properties of this matter (plasma, in my case).

So, what happened actually? In 2004, I studied the plasma properties of the boron target ablation in the Institute for Solid State Physics (Japan) using both strong (femtosecond) and relatively weak (picosecond) intensity pulses. During those days, I have analyzed the shadowgraphs of the boron and molybdenum plasma clouds. For this, I created the plasma plumes, particularly using 300 ps pulses on the boron surface and irradiated this cloud of plasma using a broad beam of second harmonic of Ti:sapphire laser ($\lambda \sim 400$ nm) [23]. The task was to measure a speed of wavefront of plasma for explanation of the three-half harmonic observed earlier in these experiments [24].

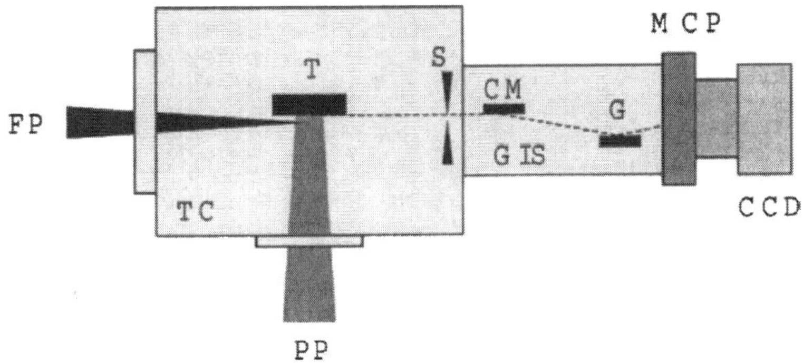

Figure 1.5 Schematic of the experimental setup for HHG: TC, target chamber; T, target; S, slit; GIS, grazing incidence spectrometer; CM, gold-coated cylindrical mirror; G, grating; MCP, micro-channel plate; CCD, charge-coupled device; FP, femtosecond pulse; PP, picosecond pulse. Reproduced from Ref. 25 with permission from Optical Society of America.

Our equipment allowed the observation of plasma spectra in the XUV range. We often analyzed the spectral characteristics of plasmas produced by pulses of picosecond and femtosecond duration. That day, I decided to check the plasma spectra after double excitation of boron plasma (Fig. 1.5) at different delays between two (picosecond and femtosecond) pulses. I produced the plasma using 300 ps pulses and then, after 40 ns, sent the focused 110 fs pulse through the plasma medium. Commonly, I overexcited the targets by the first (picosecond) pulse, to analyze a difference between the plasma spectra originated from the first pulse and after excitation of strongly ionized plasma by femtosecond pulse. This time, I decided to see what happened when the plasma becomes initially at under-excited conditions of a relatively weak interaction of the picosecond pulses with the target, without the influence of femtosecond pulse. No plasma lines, as expected, were observed in that case. However, once I sent a focused femtosecond pulse through the plasma area, an impressive picture appeared on the computer monitor, the equidistant narrow lines resembling harmonic spectral distribution (Fig. 1.6a). The increase or decrease of the fluence of heating pulse on the target surface above or below some level led to disappearance of the harmonic spectra. In the case of over-

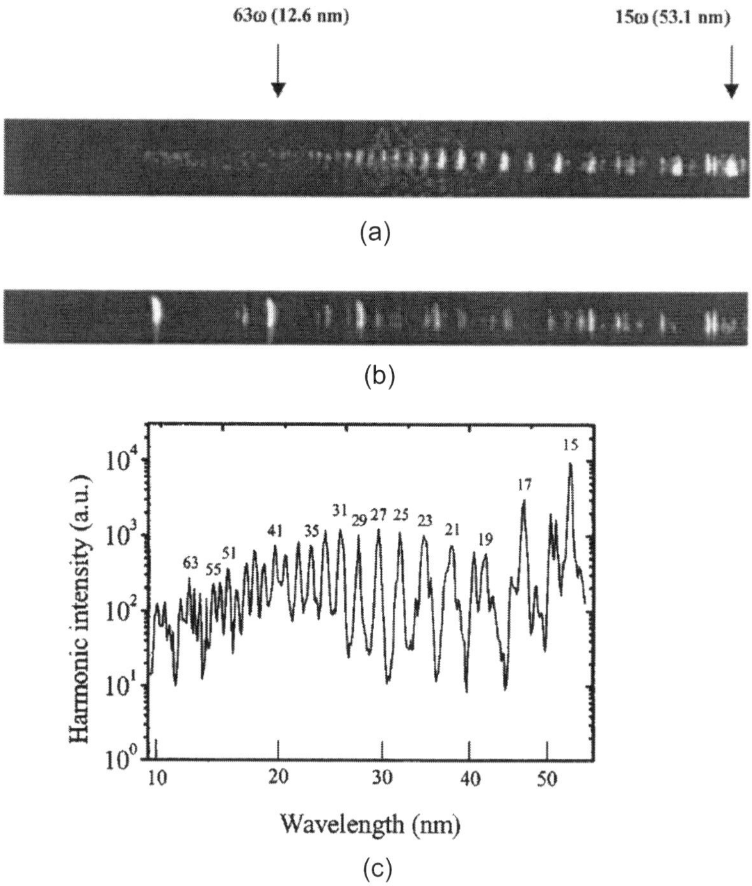

Figure 1.6 (a) Raw images of boron harmonic spectra and (b) boron ionic spectral lines at strong excitation. (c) Harmonics generated from the B plasma. Reproduced from Ref. 25 with permission from Optical Society of America.

excitation of boron target, strong ionic lines attributed to singly, doubly, and triply charged particles, as well as second and above orders of diffraction of extremely strong 6.03 and 4.86 nm emissions dominated in the XUV spectra (Fig. 1.6b). The harmonics spread along the whole area of fixed micro-channel plate. The calibration of our XUV spectrometer confirmed that these lines correspond to odd integers of the wavelength of our nearly 800 nm driving radiation.

The harmonics were extended up to the sixties orders (Fig. 1.6c). I would remind that all previous studies showed the weak harmonics from many plasma plumes, which did not exceeded twenties orders at the best conditions. In the meantime, in my case, I was able to see the images of these harmonics just directly on the phosphor screen by a naked eye, without the CCD amplifier.

Thus, the occasional observation of the high-order nonlinear optical process in plasma at specific conditions of target ablation allowed formulating main requirement for efficient HHG in this medium. One can say that it was just a luck. However, I, being involved in the studies of those harmonics at the mid of nineties, for a long time wondered whether it is possible to overcome the obstacles in generation of plasma harmonics. The first observation of the harmonics generating at new conditions of plasma formation was some sort of reward for my efforts in the studies of various properties of low-excited plasma.

We sent these results in *Optics Letters*, and they immediately accepted our paper and published it in 2005 [25]. This was a beginning of my studies of various plasma formations and their high-order nonlinear optical properties.

From then, plasma HHG has opened new doors in many unexpected areas of light-matter interaction. Apart from considering as an alternative method for generation of coherent XUV radiation, it can be used as a powerful tool for various spectroscopic and analytical applications. A few of them have already emerged during recent years of plasma HHG studies. The application of doubly charged ions for HHG showed a promising extension of the cutoff photon energy in plasma harmonics, without having to rely on few-cycle driving pulses. As it has been shown in the case of low and high-order plasma harmonics, the conversion efficiency can be strongly enhanced by making use of resonances in atomic or ionic systems. This has been demonstrated in pioneering plasma harmonic experiments using the indium and other metals. For laser-generated plasmas, a large variety of materials can be employed, thereby increasing the chance to select such resonances with fixed-frequency Ti:sapphire lasers. Furthermore, it has been shown that two-color pumping profitably enhances the high-harmonic intensity and significantly influences the output and properties of harmonic spectrum in rare gases. For plasma harmonics, where this technique has recently been adopted, this is a new approach for the nonlinear spectroscopy of ionic transitions possessing high oscillator strengths [15].

Most interestingly, recent studies have shown that enhanced high-order harmonics can also be generated from the ablated nanoparticles, which opens the prospects for applications of local field enhancement, applications of broad plasmonic resonances in the XUV range, and a more efficient recombination process for plasma HHG. As a highly interesting perspective, an increase of the harmonic output by quasi-phase matching in specially prepared plasmas may be considered. For the plasma, a different and more flexible technique than used in neutral gases can be applied. The plasma may be spatially modified using a long pulse co-propagating with the fundamental driving pulse, and conditions might be found where quasi-phase matching is possible over a long distance in the plasma, although the constructive and destructive interference in such plasmas containing different emitters can provide a new knowledge about the phase-related characteristics of this process. The alternative to this method of plasma modulation, the targets with modulated surface structures, could be used.

Thus, the above approach can be useful for producing an efficient source of short-wavelength ultra-short pulses for various applications and studies of the properties of harmonic emitters. The laser ablation induced HHG spectroscopy is a new method for the studies of material science and one of the most important applications of HHG. In the following section, I discuss the realization of new ideas appearing between 2005 and 2007, which further improved the HHG efficiency through harmonic generation in specially prepared plasmas and allowed the spectral and structural studies of matter through the plasma harmonic spectroscopy.

1.4 Improvements of Plasma Harmonics

The search for ways of increasing the notoriously low HHG efficiency in the XUV spectral range has long been (and still is) among the most topical problems of nonlinear optics. However, in the majority of cases, the efficiency of conversion to high-order harmonics turns out to be insufficient for using them as real coherent short-wavelength radiation sources in biology, plasma diagnostics, medicine, microscopy, photolithography, XUV coherent diffraction imaging, time-resolved measurements, and so on. The feasibility of increasing the intensities of high-order harmonics generated in gas jet sources by using atomic and ion resonances has been studied primarily by theoretical methods [26, 27]. The results of a number of calculations suggest that the intensity of harmonic may be substantially increased when this harmonic is at resonance with transitions in the atomic and ion spectra of gases. This approach, which has yet been realized in the gas HHG, may be an alternative (or a complement) to the method of wave phase matching for harmonics and laser radiation [28, 29].

As it was shown in Section 1.2, first experiments with plasma harmonics using over-excited plasma plumes and non-optimized conditions of laser plasma interaction led to frustrating results. Nevertheless, there was a reason to hope that harmonic intensities may be increased and efficient shorter wavelength coherent radiation may be obtained using laser-produced plasmas. There are no fundamental limitations here; it only remains to find the optimal conditions for producing a plasma plume to serve as the efficient nonlinear medium for HHG. Laser-produced plasma may be validly used for this process if the effect of the limiting factors (self-defocusing, self-phase modulation, and wave phase mismatch of the harmonics and the radiation being converted) is minimized, as it was underlined in early studies of plasma HHG [16, 20, 21].

Among the special features of HHG in laser-produced plasmas, we first of all note a wide range of nonlinear medium characteristics available by varying the conditions of laser plume production on the surface of a solid. This applies to plasma parameters such as the plasma dimension, the density of ions, electrons, and neutral particles, and the degree of their excitation. The use of any elements of the periodic table that exist as solids largely extends the range of materials employed, together with thousands of complex solid-state samples, whereas only a few light rare gases are typically used in gas HHG. Thus, the exploration of practically any available solid-state material through the nonlinear spectroscopy comprising laser ablation and harmonic generation can be considered as a new tool for materials science.

In several cases, this method furnishes an opportunity to realize quasi-resonance conditions for an increase in the efficiency of single harmonic generation due to the effect of ion transitions on the nonlinear response in the spectral range in question, thus allowing the studies of ionic transitions possessing strong oscillator strengths. This effect can hardly be observed in gas HHG because of a low probability for the coincidence of the atomic transition frequencies of few gases and the frequencies of single harmonics. The advantages of plasma HHG could largely be realized with the use of a low-excited and weakly ionized plasma, because the limiting processes governing the dynamics of laser wavelength conversion would play a minor role in this case, as it was shown in previous section in the case of boron plasma. This assumption has been confirmed by several other studies concerned with HHG in the plasma media [30–32]. A substantial increase in the highest order of the generated harmonics, the observation of long plateau and emergence of a second plateau in the energy distribution of highest order harmonics, the high efficiencies obtained with several plasma formations, the realization of resonance enhancement of individual harmonics, the efficient harmonic enhancement from plasma plumes containing clusters of different materials, and other properties revealed in those and other works [33–36] have demonstrated the advantages of using specially prepared plasmas for HHG. The orders of harmonics obtained in plasma media to date range into the sixties and seventies [25, 37, 38]. The highest order harmonics (the 101st order, wavelength 7.9 nm) have been demonstrated in manganese plasmas [39]. The HHG conversion efficiency in the plateau region amounted to 10^{-5} [40]. In addition to that, the efficiency of conversion to an individual (resonantly enhanced) high-order harmonic approached 10^{-4} [30, 36].

The quest for new plasma media that would favor the enhancement of an individual harmonic allows further enhancement of harmonic conversion efficiency. The production of a single high-intensity harmonic (rather than a group of harmonics of equal intensity in the plateau region) would open up the way to the practical application of these coherent short-wavelength radiation sources. Resonantly enhanced harmonics observed in several plasma media allowed expecting that similar conditions will be discovered for other plasma formations. The generated harmonic wavelength may then be tuned to the transitions with high oscillator strength by wavelength tuning of the driving laser [30, 35], as well as by varying the chirp of the laser radiation [32, 34]. Application of ablated nanoparticles and clusters for HHG can also enhance the yield of harmonics in the XUV range. Further improvements in HHG conversion efficiency and harmonic extension require a systematic study of the influence of various plasma and laser parameters on ablation harmonics. Many new features of plasma harmonics emerged during last few years allow expecting further extension of our knowledge of material properties using this powerful tool of nonlinear spectroscopy. Some peculiarities of those studies published after 2008 were discussed in my two recent monographs [15, 41]. In the present book, I later on concentrate on the description of the achievements in this field reported during last couple of years.

So, the answer on the question why the plasma harmonics are need at all and why they became more and more popular is presented in latest couple of paragraphs, as well as in the last paragraphs of Section 1.3, wherein the perspectives of future developments are briefly described. I think that it is more than enough to have a motivation for further development of this technique. I also hope that it became obvious that plasma HHG is not simply another method for generation of coherent XUV light, but rather a new technique for analysis of various features of harmonic emitters appearing in the plasma plumes during laser ablation of solids. Because of this, the Ockham's principle is not applicable to this new method, as it gives new knowledge, which could not be acquired by gas and surface harmonic techniques.

To push further this new method, one has to clearly understand the mechanisms of HHG in an isotropic medium, as well as peculiarities of plasmas compared with gases. In the following chapter, I will describe some main theoretical studies related with the HHG, and particularly plasma HHG.

References

1. J. Franklin, *The Science of Conjecture: Evidence and Probability before Pascal,* The Johns Hopkins University Press, 241 (2001).
2. M. Ferray, A. L'Huillier, X. F. Li, L. A. Lompré, G. Mainfray, G. Manus, *J. Phys. B: At. Mol. Opt. Phys.* 21, L31 (1988).
3. L. A. Lompré, A. L'Huillier, M. Ferray, P. Monot, G. Mainfray, G. Manus, *J. Opt. Soc. Am. B* 7, 754 (1990).
4. A. McPherson, G. Ginson, H. Jara, N. Johann, I. A. McIntyre, K. Boyer, C. K. Rhodes, *J. Opt. Soc. Am. B* 4, 595 (1987).
5. T. Desai, H. Daido, M. Suzuki, N. Sakaya, A. P. Guerreiro,K. Mima, *Laser Part. Beams,* 19, 241 (2011).
6. T. Ozaki, R. A. Ganeev, A. Ishizawa, T. Kanai, H. Kuroda, *Phys. Rev. Lett.* 89, 253902 (2002).
7. S. Suckewer, P. Jaegle, *Laser Phys. Lett.* 6, 411 (2009).
8. B. D. Patterson, R. Abela, *Phys. Chem. Chem. Phys.* 12, 5647 (2010).
9. E. Kur, D. J. Dunning, B. W. J. McNeil, J. Wurtele, A. A. Zholents, *New J. Phys.* 13, 063012 (2011).
10. C. Vozzi, F. Calegari, F. Frassetto, M. Negro, L. Poletto, G. Sansone, P. Villoresi, M. Nisoli, S. De Silvestri, S. Stagira, *Laser Phys.* 20, 1019 (2010).
11. B. Dromey, M. Zepf, A. Gopal, K. Lancaster, M. S. Wei, K. Krushelnick, M. Tatarakis, N. Vakakis, S. Moustaizis, R. Kodama, M. Tampo, C. Stoeckl, R. Clarke, H. Habara, D. Neely, S. Karsch, P. Norreys, *Nature Phys.* 2, 456 (2006).
12. G. P. Zhang, *Int. J. Modern Phys. B* 21, 5167 (2007).
13. P. Gibbon, *Short Pulse Laser Interactions with Matter: an Introduction,* Imperial College Press, London (2005).
14. U. Teubner, P. Gibbon, *Rev. Modern Phys.* 81, 445 (2009).
15. R. A. Ganeev, *High-order harmonic generation in laser plasma plumes,* Imperial College Press, London (2012).
16. Y. Akiyama, K. Midorikawa, Y. Matsunawa, Y. Nagata, M. Obara, H. Tashiro, K. Toyoda, *Phys. Rev. Lett.* 69, 2176 (1992).
17. S. Kubodera, Y. Nagata, Y. Akiyama, K. Midorikawa, M. Obara, H. Tashiro, K. Toyoda, *Phys. Rev. A* 48, 4576 (1993).
18. C.-G. Wahlström, S. Borgström, J. Larsson, S.-G. Pettersson, *Phys. Rev. A* 51, 585 (1995).
19. W. Theobald, C. Wülker, F. R. Schäfer, B. N. Chichkov, *Opt. Commun.* 120, 177 (1995).

20. R. A. Ganeev, V. I. Redkorechev, T. Usmanov, *Opt. Commun.* 135, 251 (1997).
21. K. Krushelnick, W. Tighe, S. Suckewer, *J. Opt. Soc. Am. B* 14, 1687 (1997).
22. R. A. Ganeev, *Laser Phys. Lett.* 9, 175 (2012).
23. R. A. Ganeev, M. Suzuki, M. Baba. H. Kuroda, *Opt. Spectrosc.* 99, 1000 (2005).
24. R. A. Ganeev, M. Suzuki, M. Baba, M. Turu, H. Kuroda, *Appl. Phys. B* 78, 79 (2004).
25. R. Ganeev, M. Suzuki, M. Baba, H. Kuroda, T. Ozaki, *Opt. Lett.* 30, 768 (2005).
26. F. M. C. Faria, R. Kopold, W. Becker, J. M. Rost, *Phys. Rev. A* 65, 023404 (2002).
27. R. Taieb, V. Veniard, J. Wassaf, A. Maquet, *Phys. Rev. A* 68, 033403 (2003).
28. E. A. Gibson, A. Paul, N. Wagner, R. Tobey, D. Gaudiosi, S. Backus, I. P. Christov, A. Aquila, E. M. Gullikson, D. T. Attwood, M. M. Murnane, H. C. Kapteyn, *Science* 302, 95 (2003).
29. S. Kazamias, D. Douillet, F. Weihe, C. Valentin, A. Rousse, S. Sebban, G. Grillon, F. Augé, D. Hulin, Ph. Balcou, *Phys. Rev. Lett.* 90, 193901 (2003).
30. R. A. Ganeev, M. Suzuki, T. Ozaki, M. Baba, H. Kuroda, *Opt. Lett.* 31, 1699 (2006).
31. R. A. Ganeev, M. Baba, M. Suzuki, H. Kuroda, *J. Appl. Phys.* 99, 103303 (2006).
32. R. A. Ganeev, P. A. Naik, H. Singhal, J. A. Chakera, P. D. Gupta, *Opt. Lett.* 32, 65 (2007).
33. R. A. Ganeev, M. Suzuki, M. Baba, H. Kuroda, *Appl. Phys. B* 81, 1081 (2005).
34. R. A. Ganeev, H. Singhal, P. A. Naik, V. Arora, U. Chakravarty, J. A. Chakera, R. A. Khan, P. V. Redkin, M. Raghuramaiah, P. D. Gupta, *J. Opt. Soc. Am. B* 23, 2535 (2006).
35. M. Suzuki, M. Baba, R. Ganeev, H. Kuroda, T. Ozaki, *Opt. Lett.* 31, 3306 (2006).
36. R. A. Ganeev, H. Singhal, P. A. Naik, V. Arora, U. Chakravarty, J. A. Chakera, R. A. Khan, I. A. Kulagin, P. V. Redkin, M. Raghuramaiah, P. D. Gupta, *Phys. Rev. A* 74, 063824 (2006).
37. R. A. Ganeev, H. Singhal, P. A. Naik, U. Chakravarty, V. Arora, J. A. Chakera, R. A. Khan, M. Raghuramaiah, S. R. Kumbhare, R. P. Kushwaha, P. D. Gupta, *Appl. Phys. B* 87, 243 (2007).

38. M. Suzuki, M. Baba, H. Kuroda, R. A. Ganeev, L. B. Elouga Bom, T. Ozaki, *Opt. Express* 15, 4112 (2007).
39. R. A. Ganeev, L. B. Elouga Bom, J.-C. Kieffer, T. Ozaki, *Phys. Rev. A*, 76 023831 (2007).
40. R. A. Ganeev, M. Baba, M. Suzuki, H. Kuroda, *Phys. Lett. A*, 339 103 (2005).
41. R. A. Ganeev, *Nonlinear optical properties of materials*, Springer Series in Optical Sciences, Vol. 174 (2013).

Chapter 2

Theoretical Basics of Plasma Harmonics

2.1 Basics of HHG

HHG is a nonlinear optical process that converts light of one frequency, called the fundamental or driving, into light at integer multiples of that fundamental frequency, called the harmonics. The observation of harmonics can be considered as having two parts: their generation at the single atom level and their propagation to coherently build up a macroscopic signal over a number of emitters. Below, we consider both low-order and high-order harmonic generation [1].

Low-order (second or third) harmonic generation in solid materials has had huge technological impacts in laser technology, such as green laser pointers or the pump lasers used in Ti:sapphire lasers. The third harmonic generated in gases was first observed by Ward and New in 1969 [2]. Since then, the many advances in laser technology and the keen interest by a fast growing community has extended HHG to the generation of photons of several hundred eVs of energy [3–7].

If a single atom is tunnel ionized by an intense laser pulse, its outer-most electron will appear in the continuum at rest in the presence of an oscillating electric field. The phase of the field at the

Plasma Harmonics
Rashid A. Ganeev

ISBN 978-981-4463-30-0 (Hardcover), 978-981-4463-31-7 (eBook)
www.panstanford.com

time at which the electron appears in the continuum is what will determine its destiny. But first let us consider the generation of low-order harmonics, where the emitted photon energy is less than or around the value of the binding potential of the outer-most electron. In the photon picture, the outer-most electron absorbs a laser photon and moves up to a virtual state further absorbing photons from there if the photon density is great enough that it has a significant probability to do so before it has time to decay back down to the ground state. Eventually, it does decay back down, emitting a photon whose energy is the sum of the energies of the absorbed photons.

Here, the low-order perturbation theory is still a fast and accurate way to calculate the harmonic emission. The polarization induced in the medium, P, significantly deviates from its linear response [8]:

$$P = \epsilon_0 \sum_{n=1}^{N} \chi^{(n)} E^n \tag{2.1}$$

The linear response is the case for which $N = 1$ and the xth order harmonic is generated for $N = x$. For centro-symmetric media, all the $x^{(n)}$ for even n are equal to zero. The low-order perturbation theory is convenient for low-order harmonics but becomes very computationally demanding as the order increases.

When the ponderomotive potential becomes comparable or greater than ionization potential I_p, it becomes easier to think of the process as non-perturbative. This is the most common way of looking at HHG. A semi-classical theory was first developed by Corkum [9] and then a fully quantum mechanical model was presented by Lewenstein et al. [10].

The most heuristic way to describe HHG that retains a physical basis is using the semi-classical three-step model (Fig. 2.1 [11]). The first step is the ionization of the atom by the laser field, the second is the acceleration of the electron by the laser field, and the third is the recombination with its parent ion, which can lead to the emission of a harmonic photon [12]. In the wave function picture, this can be seen as the original wave function of the ground state electron, Ψ_0, being split into a bound part, Ψ_b, which can be regarded as a Gaussian function, and a continuum part, Ψ_c, which can be treated as a plane wave. This means that if and when Ψ_c returns to the same location as Ψ_b, they generate an oscillating

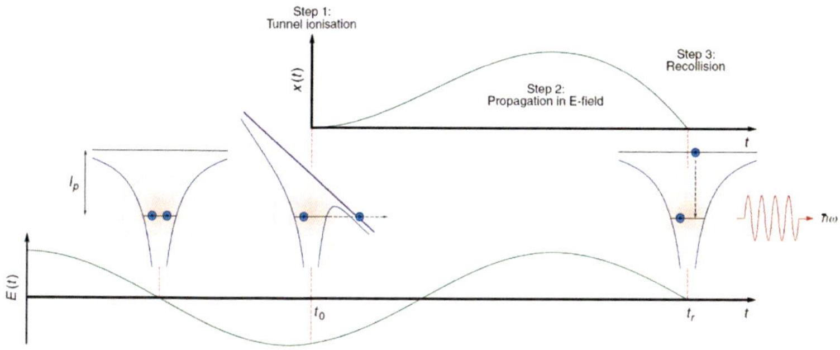

Figure 2.1 Three-step mechanism of high-order harmonic generation: step 1–tunnel ionization, step 2–electron acceleration in the electromagnetic field of the laser wave, and step 3–recollision and recombination with the ion and the emission of harmonics. Reproduced from Ref. 11 with permission from IOP Publishing. All rights reserved.

dipole and the acceleration of this dipole can be observed through the harmonic spectrum. Furthermore, it means that the continuum electron can only recombine with the ground state, which makes HHG a unique tool, as the recombination process is state selective and the spectrum does not retain information about any other channel.

The strong field approximation (SFA) can be used to simplify the calculations of the harmonic spectrum. The main approximations associated with it are that the bound electron does not feel the influence of the external electric field and that the continuum electron does not feel the influence of the ionic potential. This is justified by the fact that the ionization is dominated by tunnelling; it is therefore peaked around the time of maximum laser electric field. This means that the field is strong at the time when the electron appears in the continuum and the motion of the ionized electron is widely dominated by the laser field. When the laser field reaches zero, the electron is far from the ion core and this assumption is still valid. When the electron returns, its velocity is very high so the time the ionic potential has to affect the electron's trajectory is short and it is still a reasonable approximation to discard it. This is a semi-classical model in the way that the evolution of the electron in the continuum is treated classically and the ionization and recombination are treated quantum mechanically.

The validity of the SFA has been tested against experimental results many times and although having to ensure that the experiment is done within its range of validity, it has proven that it can be a very useful and accurate way of calculating harmonic spectra [13, 14]. The more thorough way of calculating the spectra can be done by solving the time-dependent Schrödinger equation (TDSE). This method is a lot more computationally demanding, as no approximations apart from the form of the ground state are made. It also gives less insight into the underlying physics, as there is no distinction between all the factors, which affect the generation process. The intermediate step between the TDSE and SFA calculations is the Lewenstein model [10], which remains quantum mechanical but still ignores the effect of the ionic potential on the continuum electron and only considers one active electron. The wave function, $\Psi(\mathrm{t})$, can be described by:

$$|\Psi(t)\rangle = e^{iIpt}\left(a(t)|0\rangle + \int d^3\mathrm{v}b(\mathrm{v},t)|\mathrm{v}\rangle\right) \tag{2.2}$$

The oscillations of the ground state phase e^{iIpt} are factored out at the front and $a(t)$ and $b(\mathrm{v,t})$ are the amplitudes of the ground and continuum states with $\mathrm{D}(t) = \langle\Psi(t)|\hat{\mathrm{r}}|\Psi(t)\rangle$ velocity v. The dipole moment can be calculated as:

$$\begin{aligned}\mathrm{D}(t) = i\int_0^t dt' \int d^3\mathrm{p}a^*(t)\mathrm{d}^*(\mathrm{p}-\mathrm{A})(t')\\ \times a(t')\mathrm{E}(t')\cdot\mathrm{d}(\mathrm{p}-\mathrm{A})(t)e^{-iS(\mathrm{p},t,t')} + c.c.\end{aligned} \tag{2.3}$$

The dipole matrix element for the bound-free transitions, $\mathrm{A}(t)$, is $\mathrm{d}(\mathrm{v}) = \langle\mathrm{v}|\hat{\mathrm{r}}|0\rangle$ vector potential associated with the electric field $\mathrm{E}(t)$; $\mathrm{p} = v + \mathrm{A}(t)$ is the canonical momentum, and the semi-classical action, S, describes the motion of the electron in the continuum:

$$S(p,t,t') = \int_{t'}^{t}\left(\frac{[\mathrm{p}+\mathrm{A}(t')]^2}{2} + I_p\right) \tag{2.4}$$

The dipole moment can be twice differentiated with respect to time to give the dipole acceleration. This dipole acceleration can be Fourier transformed into the harmonic spectrum.

All the above-mentioned models agree on the shape of the harmonic spectrum if used in their range of validity. TDSE sees an exponentially decaying region for the low-order harmonics where the interaction can be treated as perturbative. The SFA results for that part of the spectrum do not make any sense, as the influence of the ionic potential cannot be ignored. All models then see a long plateau where the intensity stays around the same order of magnitude for many harmonic orders followed by a sharp cutoff. The SFA calculations locate the cutoff energy, E_{co}, to be at:

$$E_{\text{co}} = I_{\text{P}} + 3.17U_{\text{P}} \tag{2.5}$$

where I_{P} is the ionization potential of the atom and U_{P} is the ponderomotive energy.

The full quantum mechanical description yields a slightly different cutoff energy, E_{coq}:

$$E_{\text{coq}} = I_{\text{P}}F(I_{\text{P}}/U_{\text{P}}) + 3.17U_{\text{P}} \tag{2.6}$$

where the factor $F(I_{\text{P}}/U_{\text{P}})$ is equal to 1.3 for $U_{\text{P}} \gg I_{\text{P}}$ and tends to one as U_{P} decreases [12]. A higher ionization potential of the chosen target leads to a greater cutoff harmonic order. The use of low-frequency fields results in higher cut-off energies, as the electron spends more time in the field before recollision, building up more energy. This is counterbalanced by a very rapid loss of harmonic conversion efficiency for longer wavelengths, the efficiency scaling being around λ^{-5} [15].

The cutoff in the harmonic spectrum can be a useful tool to work out the peak intensity of the interaction. Care must be taken that the harmonics are generated well under the saturation of the ionization of the generating medium, as this would lead to an underestimate of the intensity.

High harmonics have a number of interesting properties. They are a tunable table-top source of XUV/soft X-rays, synchronized with the driving laser and produced with the same repetition rate. The harmonic cutoff varies linearly with increasing laser intensity up until the saturation intensity I_{sat} where harmonic generation stops. The saturation intensity can be increased by changing the atomic species to lighter noble gases, but these have a lower conversion efficiency so there is a balance to be found depending on the photon energies required.

High harmonic generation strongly depends on the driving laser field, and as a result, the harmonics have similar temporal and spatial coherence properties. High harmonics are often generated with pulse durations shorter than that of the driving laser. This is due to phase matching and ionization. Often harmonics are only produced in a very small temporal window when the phase matching condition is met. Depletion of the generating media due to ionization also means that harmonic generation is mainly confined to the leading edge of the driving pulse. High harmonics are emitted co-linearly with the driving laser and can have a very tight angular confinement, sometimes with less divergence than that of the fundamental field and near Gaussian beam profiles.

Some interesting limits on the HHG process, which are explained by recollisional model, show that HHG will only occur if the driving laser field is linearly polarized. Ellipticity on the laser beam causes the returning electron to miss the parent nucleus. Quantum mechanically, the overlap of the returning electron wavepacket with the nuclear wavepacket is reduced. This has been observed experimentally, where the intensity of harmonics decreases rapidly with increasing ellipticity. Another effect, which limits the intensity of the driving laser, is the Lorentz force. At intensities above 10^{16} W/cm^2, the magnetic component of the laser pulse, which is ignored in weak field optics, can become strong enough to deflect the returning electron. This will cause it to "miss" the parent nucleus and hence prevent HHG.

Now, once we have discussed the basics of harmonics generation in isotropic medium, let us consider some theories describing HHG in various media produced by laser ablation. Particularly, I present for your consideration the analysis of the HHG in fullerenes, discuss various approaches for description of observed peculiarities of resonant enhancement of a single harmonic in laser plasma, show the examples of theoretical description of the two-color pump resonance-induced enhancement of odd and even harmonics from a tin plasma, and present the calculations of single harmonic generation of few-cycle pulses from Mn plasma.

2.2 Harmonic Generation in Fullerenes Using Few-Cycle Pulses

Fullerenes can be considered as an attractive nonlinear medium for the HHG. Their relatively large sizes and broadband surface plasmon resonance (SPR) in the extreme ultraviolet range allowed the first demonstration of efficient HHG from fullerenes near their SPR ($\lambda_{SPR} \approx 60$ nm, with 10 nm full width at half maximum) [16]. The application of laser ablation technique led to the creation of dense C_{60}-rich plasma ($\sim 5 \times 10^{16}$ cm^{-3}), in a stark contrast with the density less than 10^{13} cm^{-3} obtained using the heat oven based methods for production of the fullerene beams.

Theoretical studies of HHG from C_{60} using the multi-cycle pulses involved extending the three-step model [17], analyzing an electron constrained over the surface of a rigid sphere, with geometrical parameters similar to those of the C_{60} fullerene [18], and using the dynamical simulations [19]. In the latter work, higher order harmonics were shown to be due to multiple excitations and could be easily generated even with a weak laser field. Those studies reveal how HHG can be used to probe the electronic and molecular structure of C_{60}. At the same time, theoretical investigation of such systems is hampered by the fact that the Hamiltonian of HHG is time dependent and the systems consist of many electrons. The investigation of the influence of the fundamental properties of electrons on the resonant HHG can be performed by means of a multiconfigurational time-dependent Hartree–Fock (MCTDHF) approach, which has the accuracy of direct numerical solution of Schrödinger equation and is almost as simple as the ordinary time-dependent Hartree–Fock (TDHF) approach. In particular, the computations could be based on the Heidelberg multicongurational time-dependent Hartree (MCTDH) software packages [20]. In [21], simulations of resonant HHG are performed by means of a MCTDHF approach for three-dimensional fullerene-like systems. The influence of the SPR of C_{60} on the harmonic efficiency in the range of 60 nm ($E = 20$ eV) was analyzed and showed the ways of resonant HHG optimization.

The saturation intensities of different charge states of C_{60} are higher than isolated atoms of similar ionization potential [22]. In this connection, it would be interesting to analyze the behavior of fullerene molecules in the field of few-cycle pulses from the point of view of the harmonic generation and compare these studies with those carried out using the multi-cycle pulses.

Once considered exotic, fullerenes are nowadays commonly synthesized and used in research. Fullerenes have a graphite-like structure, but instead of purely hexagonal packing, they also contain pentagons (or even heptagons) of carbon atoms, which bend the graphene sheet into spheres, ellipses, or cylinders. Among fullerenes, C_{60} is particularly noteworthy, as its pentagonal and hexagonal rings of carbon atoms give it a spherical shape. This feature endows C_{60} with an appealing spherical symmetry, which can be used to simplify the treatment [23]. Here, we briefly outline the theory of a C_{60} molecule driven by a laser pulse of the form $E(t) = E_0(t)\sin(\omega_L t)$ polarized in the z direction with $E_0(\mathrm{t})$ describing the on-off switching of the pulse profile and frequency ω_L; in the present calculations, the pulse profile has been always taken as trapezoidal with two optical cycles of switching on and off [24].

The C_{60}–laser interaction will be described in the single active electron approximation with one electron constrained over a structureless, spherical surface of radius $R = 3.55 \times 10^{-8}$ cm $= 6.71a_0$; the electron is otherwise unconstrained.

Within the model, the Hamiltonian $\hat{H}$ of the problem is:

$$\hat{H} = \hat{H}_0 + \hbar\Omega_0(t)\cos\theta\sin(\omega_L t) \tag{2.7}$$

where

$$\hat{H}_0 = \frac{\hbar^2}{2L}\hat{L}^2 \tag{2.8}$$

is the Hamiltonian of the C_{60} in the absence of laser field, $\hbar\Omega_0(t) = eRE_0(t)$, $I = m_e R$ is the moment of inertia of the electron, and $\hat{L}^2$ is the square angular momentum operator whose eigenstates are the usual spherical harmonics $Y_{\mathrm{lm}}(\theta,\varphi) \to |\ell, m\rangle$: $\hat{L}^2|\ell, m\rangle = \ell(\ell+1)|\ell, m\rangle$. Nevertheless, the fact that φ is a cyclic variable makes m a constant quantum number. φ is the angle of the position of the electron with respect to the z axis. Accordingly, the energies of the bare states are

$$\hbar\omega_t \equiv +\frac{\hbar^2}{2L}\ell(\ell+1) \tag{2.9}$$

thus the gap between contiguous energy levels is linearly increasing with ℓ; this particular feature can be reminiscent of the ionization suppression of the C_{60} that results from the experimental fact that the molecule can absorb from the laser field an energy amount quite larger than the minimum required for ionization.

The time-dependent Schrödinger equation to be solved is:

$$i\hbar\frac{\partial}{\partial t}|\Psi(t)\rangle = \hat{H}\,|\Psi(t)\rangle \tag{2.10}$$

with $|\Psi(t)\rangle$ the full time-dependent wave function of the active electron acted upon by the laser.

In accordance with previous assumptions, it is convenient to write the state vector at time t, $|\Psi\,(t)\rangle$, as a linear combination of eigenstates of the laser-free Hamiltonian $\hat{H}_0$:

$$|\Psi(t)\rangle = \sum_{\ell'=0}^{\infty}\sum_{m'=-t}^{\ell} a_{\ell',m'}(t)|\ell,' m'\rangle \tag{2.11}$$

where $a_{\ell',m'}(t)$ are expansion coefficients to be found by substitution into the time-dependent Schrödinger equation. Let $|\Lambda, m\rangle$ be the initial state assumed to be the highest occupied molecular orbital; accordingly to the previous discussion, the laser will couple only states with $m = m'$. Therefore, the time-dependent Schrödinger equation becomes

$$i\dot{a}_{\ell,m} = \omega_\ell a_{\ell,m} + \Omega_0 \sum_{\ell'=\Lambda}^{\infty} \sin(\omega_{\mathrm{L}}t)\langle\ell, m|\cos\theta|\ell', m\rangle a_{\ell',m} \tag{2.12}$$

In fulfillment of the Pauli exclusion principle and of the single active electron approximation, the summation does not extend over occupied core states. Thus, the matrix element reduces to a standard integration of three spherical harmonics:

$$\langle\ell, m|\cos\theta|\ell', m\rangle = \frac{4\pi}{3}\int Y^*_{\ell,m}(\theta,\varphi)Y_{1,0}(\theta,\varphi)Y_{\ell,m}(\theta,\varphi)d\Omega \tag{2.13}$$

By setting

$$b_{\ell,m} \equiv \sqrt{\frac{(\ell+m+1)(\ell-m+1)}{(2\ell+1)(2\ell+3)}} \tag{2.14}$$

one can obtain

$$\begin{aligned} &\int Y^*_{\ell,0}(\theta,\varphi)Y_{1,0}(\theta,\varphi)Y_{\ell+1,m}(\theta,\varphi)d\Omega = \sqrt{\frac{3}{4\pi}}b_{\ell,m} \qquad \text{If } \ell \geq \Lambda \\ &\int Y^*_{\ell,0}(\theta,\varphi)Y_{1,0}(\theta,\varphi)Y_{\ell-1,m}(\theta,\varphi)d\Omega = \sqrt{\frac{3}{4\pi}}b_{\ell-1,m} \qquad \text{If } \ell > \Lambda \end{aligned} \tag{2.15}$$

Therefore,

$$\begin{aligned} i\dot{a}_{\Lambda,m} &= \omega_\Lambda a_{\Lambda,m} + \Omega_0(t) b_{\Lambda,m} \sin(\omega_L t) a_{\Lambda+1,m} \\ i\dot{a}_{\ell>\Lambda,m} &= \omega_\ell a_{\ell,m} + \Omega_0(t) \sin(\omega_L t)(b_{\ell-1,m} a_{\ell-1,m} + b_{\ell,m} a_{\ell+1,m}) \end{aligned} \quad (2.16)$$

As the energy gap increases with ℓ, one can argue that the population of the upper levels becomes negligible, thus the energy levels with quantum number $\ell > \Lambda + N$ do not participate in the dynamics of the system. The set of differential equations shows a ladder-like structure so that any energy level is coupled to the two nearest levels. This system can be numerically solved by use of the matlab numerical routine odesolve45.

Once the probability amplitudes $a_{k,m}$ have been obtained, it is simple to get the electric dipole moment $er(t)$ induced by the laser on C_{60} with:

$$\begin{aligned} r(t) &= \langle \Psi(t) | R \cos\theta | \Psi(t) \rangle \\ &= R \sum_{s=0}^{\infty} b_{\Lambda+s,m} [a^*_{\Lambda+s,m} a_{\Lambda+s+1,m} + a_{\Lambda+s,m} a^*_{\Lambda+s+1,m}] \end{aligned} \quad (2.17)$$

The calculated harmonic spectra from C_{60} are presented in Fig. 2.2 in the case of 780 nm (photon energy $E_{\text{ph}} = 1.6$ eV) and 1300 nm (photon energy $E_{\text{ph}} = 0.96$ eV) pulses propagating through the fullerene medium [24]. The calculations were carried out for two-cycle pulses ($t = 5.2$ fs) and 12-cycle pulses ($t = 31$ fs) of 780 nm radiation and eight-cycle pulses ($t = 34$ fs) of 1300 nm radiation and intensity 6×10^{14} W/cm^2, which was close to the experimental conditions of fullerene HHG [16, 24–26].

The spectra are formed of well-resolved harmonics but with broadened lines (in the case of short pulses) and hyper Raman lines (in the case of long pulses). Hyper Raman lines with frequency other than harmonics are due to transitions between laser-dressed molecular states [27, 28]. The presence of these lines has been predicted since the very beginning of the theoretical treatment of HHG [29] and found in different contexts such as two-level approximation, quantum dots calculations, hydrogen atom, and so on [30–33] but never observed in actual experiments. Several explanations have been proposed to explain this failure; for example,

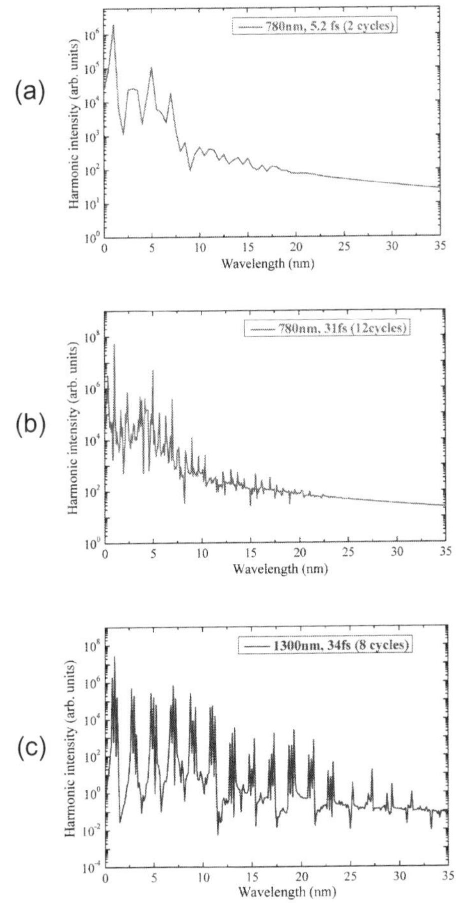

Figure 2.2 Calculated data of the harmonic spectra from fullerene plasma in the case of (a,b) 780 nm and (c) 1300 nm probe radiation. The pulse durations are (a) 2 optical cycles (5.2 fs), (b) 12 optical cycles (31 fs), and (c) 8 optical cycles (34 fs). Reproduced from Ref. 24 with permission from Optical Society of America.

it has been argued that they add destructively in the forward direction or that they show a transient nature and are thus overwhelmed by the presented odd harmonics [34, 35].

The calculations showed well-defined harmonics (up to $H_c = 31$) in the case of 1300 nm multi-cycle pulses. This theoretical model exploited the spherical symmetry of C_{60} by introducing a number of approximations, the most important of which is that the molecule cannot be ionized. This approximation deserves some comment. In spite of the ionization suppression of the C_{60} molecule, some ionization is bound to occur so that the theory becomes unreliable when ionization becomes significant. The harmonic cutoff in the case of 1300 nm radiation was extended compared with 780 nm radiation ($H_c = 17$), analogously as in the case of experiment [24].

2.3 Various Approaches for Description of Observed Peculiarities of Resonant Enhancement of a Single Harmonic in Laser Plasma

The dependence of the recombination probability on the electron return energy and on the structure of the target is reflected in the HHG spectrum and has been a subject of intensive research in recent years. To enhance the notoriously low efficiency of the HHG process, it appears promising to exploit the effect of resonances, which is known to be of great importance in photoionization. The investigation of resonant peaks in the photoionization cross-section has a long history, including studies of autoionizing resonances [36, 37], shape resonances [38] and giant resonances [39], but there have been only a few studies on the role of resonances in HHG.

The role of atomic resonances in increasing the laser radiation conversion efficiency was actively discussed in the framework of perturbation theory at the early stages of the study of low-order harmonic generation (see monograph [40] and the references therein). In the case of HHG, the increase in the efficiency of generated harmonics due to resonance processes came under discussion almost a decade ago, and this approach appears to have considerable promise with the use of ionic and, in some cases, atomic resonances [41–47]. These papers comprise both the theoretical treatment of the process and the description of the first attempts to form resonance conditions in gas HHG experiments. In particular, in [41], they investigated resonance-enhanced HHG, with an emphasis on the physical mechanism of the enhancement. By truncating a long-range potential, the researchers studied the significance of the long-range tail, the Rydberg series and the existence of highly excited states for the enhancements of harmonics.

Although theoretical estimates testified the possibility of an efficient enhancement of individual harmonics and groups of harmonics, experimental works revealed the difficulties encountered in observation of resonantly enhanced HHG in gases. Further developments have revealed that the use of plasma media could largely facilitate the solution to the problem of resonance harmonic enhancement. Examination of a large group of potential targets allowed identification of some of them as suitable for demonstrating this process [48, 49]. The advantages of "plasma HHG" over "gas HHG" were amply manifested in those studies, because the number of possible media in the former case is far greater than in the latter case.

Some attempts at explanation of the experimental observations of resonant enhancement in plasma harmonics have been reported recently [21, 50–56]. In particular, in [52] it has been shown that the influence of atomic autoionizing states on the phase matching of HHG may result in efficient selection of the single harmonic in calcium plasma. This was the first report of efficient high-order harmonic selection using autoionizing states. The calculations [52] show that the achievement of phase matching for the HHG of Ti:sapphire laser radiation in Ca^{+} plasma results in the selection of a single (21st) harmonic with a conversion efficiency of nearly 10^{-3}. The variations of the plasma components and fundamental wavelength result in the tuning of a selected harmonic frequency in the plateau region. The influence of ac Stark shift and free electrons changes the phase mismatch and the optimal laser frequency at which the efficient selection of a single harmonic is achieved. So, the intensity enhancement due to propagation effects can be even greater than that in the single-atom approximation.

An approach that suggests a HHG model describing enhancement of the generation efficiency for the harmonic resonant with the transition between the ground and the auto-ionizing state of the generating ion was developed by Strelkov [53]. In his model, the third (recombination) step of the three-step scenario of HHG [9] is partitioned into two steps: the capture of a laser-accelerated electron

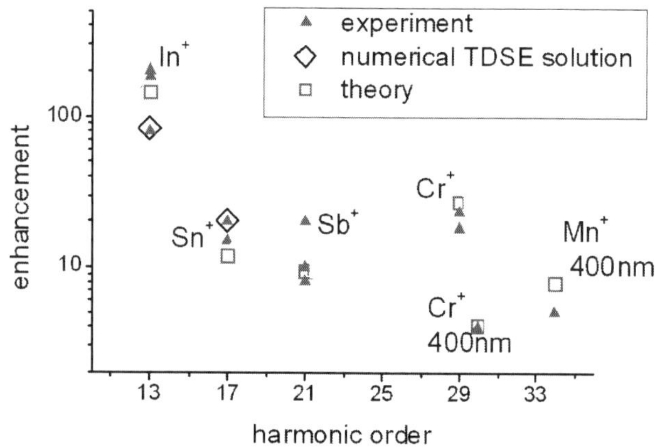

Figure 2.3 Comparison of the experimental results on resonance enhancement of harmonics and the analytical and numerical results using time-dependent Schrödinger equation solutions. Reproduced from Ref. 53 with permission from American Physical Society. Copyright 2010.

into an autoionizing state of the parent ion followed by the radiative relaxation of this state to the ground state with emission of the harmonic photon. Figure 2.3 shows calculations from [53] indicating that although the enhancement values for different media differ almost two orders of magnitude, the theoretical results are close to the experimental ones. The difference between them is attributable to the medium effects (harmonic absorption and detuning from the HHG phase matching) that are not taken into account in this theory.

Although this four-step model provides reasonable estimates for the ratio of the enhanced harmonic intensity to the averaged intensity of neighboring harmonics, the authors of [55] point out that the suppression of harmonics preceding the resonant one remains a puzzle for the theoretical model [53]. In their research [55], Frolov et al show that enhancements of single harmonics with harmonic energies near the energies of auto-ionizing states in atoms or atomic ions, as well as the aforementioned suppression of the preceding harmonics, may be interpreted (at least for those harmonics in the region of the classical plateau cutoffs for a given laser frequency and intensity) in terms of the usual three-step scenario of the HHG [9], without any additional assumptions. In particular, they successfully reproduced the main features observed in experiments on the HHG from the plasmas produced by the laser ablation of solid Cr and Mn targets. For both 800 and 400 nm wavelengths, these features are caused by atomic structure effects in the radiative recombination cross-sections of Cr^{2+} and Mn^{2+} ions (or, equivalently, in the photoionization cross-sections of Cr^{+} and Mn^{+} ions). These effects were predicted by the factorization formula [57] for the HHG rates. The experimental measurements of these rates serve to complement the measurements of Cooper minima in HHG from neutral atoms [58, 59].

In contrast to the Cooper minima in HHG, however, which have a predominantly single electron origin, the plasma HHG results can be regarded as the first experimental evidence of genuine multielectron atomic dynamics. As this dynamics cannot be accurately described by the commonly used single-active electron approximation of the time-dependent Schrödinger equation for HHG, the search for the atomic targets permitting enhanced harmonic generation in the XUV range should stimulate accurate multielectron calculations of the photoionization cross-sections from outer and sub-shells of complex atoms and ions in order to provide an accurate description of the strongly resonant transitions that underlie the enhanced yield of particular high-order harmonics.

It was found by Milošević [54] that the laser intensity dependence of the intensity and phase of the single harmonic generated in the resonant HHG from plasma ablation is different than that of the standard plateau and cutoff high harmonics. The resonant harmonic intensity increases continuously (i.e., without rapid oscillations) with the increase of the laser intensity, while the resonant harmonic phase is almost constant. Such unusual behavior of the harmonic phase requires a detailed experimental investigation.

The sub-femtosecond light pulses can be obtained by superposing several high-order harmonics. In the context of the recent first attosecond pulse train reconstruction of high-order harmonics from laser ablation plasma [60], the results of [54] are even more important. It was found that the temporal profile of a group of odd harmonics, which encompasses the resonant harmonic, is in the form of a broad peak in each laser-field half-cycle. This is an advantage in comparison with the usual plateau and cutoff harmonics wherein two such peaks are generated per half-cycle, which requires the appropriate experimental technique (i.e., such focusing that the collective effects due to the macroscopic propagation select only one peak). Taking into account a smooth dependence of the harmonic intensity on the laser intensity, and that it is not necessary to manipulate with long and short orbits by appropriate focusing, one can expect that resonant HHG has a bright perspective for applications in attoscience [61, 62].

The HHG in the presence of a shape resonance was analyzed in [56]. To understand the HHG mechanism, the time–frequency analysis of the intensity and phase was investigated. It was found that the resonance gives rise to a clear signature in the HHG spectrum irrespective of the pulse length. The time–frequency analysis supports Strelkov's four-step model, according to which the recombination process consists of two steps: capture of the returning electron into the resonance and subsequent radiative transition to the ground state. Although the one-dimensional calculation favors capture from the long trajectory of accelerated electron, one can expect that a full three-dimensional calculation will show a similar mechanism, but with higher weight given to the short trajectory of this electron. By the nature of this process, the emitted harmonic radiation is phase-locked with the usual harmonic emission from the short and long trajectories. For long-lived resonances, the interference occurs between the populations caused by re-collisions in different half cycles. This leads to new possibilities for XUV pulse shaping in the sub-femtosecond time scale.

In [21], simulations of resonant HHG were performed by means of a multiconfigurational time-dependent Hartree–Fock approach for 3D fullerene-like systems and the influence of the SPR of C_{60} on the harmonic efficiency in the range of 60 nm (E ~20 eV) was analyzed. Those results showed ways of resonant HHG optimization and, most important, attosecond pulse train generation. The MCTDHF calculations of the HHG from C_{60} clusters were in good qualitative agreement with the experimental data reported in previous studies of harmonic generation in fullerene-containing laser plumes. The broadness of the SPR in C_{60} allows direct stimulated transition from the continuum into the ground state without the additional need of radiationless transition, thus making possible competing enhancement of neighboring harmonics, which is useful for attosecond pulse train generation.

In general, multielectron SPR of C_{60} is a generalization to two-electron autoionizing states in atoms and simple molecules. However, the extreme width of the plasmon resonance allows direct recombination, whereas for auto-ionising states radiationless transition to these states should happen first. Usage of strongly ionized medium with some delocalized electrons as a target for resonant HHG can be favorable for extension of such attosecond pulse trains into the water window spectral region.

The work [63] was devoted to an explanation of the observed phenomena of resonant enhancement of a single harmonic in indium plasma without enhancement of neighboring ones. It can also be used to predict the most promising targets for resonant HHG and to increase its efficiency by control of the pump radiation's parameters. All computations by Redkin et al. [63] were performed using the time-dependent density functional theory (TDDFT) [64] with the aid of real-space real-time code OCTOPUS [65, 66], which is a software package for performing Kohn–Sham TDDFT calculations. A detailed description of the TDDFT formalism can be found in [67]. The results of TDDFT calculations of the HHG in indium were found to be in good qualitative correlation with the experimental data. This allowed proof of the possibility of direct resonant recombination for HHG on the basis of calculations for artificially chirped pulses. The peculiarities of this approach were discussed and compared with existing theories of resonant enhancement of harmonics.

As a general conclusion, in any time-dependent HHG calculation, which supports strong excited states, a resonant HHG should be observed if resonant conditions are met at the moment of recombination. The states themselves can be artificially introduced for single-electron models [59] or can follow naturally from the potential well structure in multielectronic calculations. The examples of such approach are presented in the following subsections.

2.4 Two-Color Pump Resonance-Induced Enhancement of Odd and Even Harmonics from a Tin Plasma

Among a few laser-produced plasmas demonstrating enhanced harmonics, tin represents an interesting sample of single harmonic generation, when strong transitions of singly and doubly charged ions can considerably influence this process depending on the experimental conditions (wavelength of driving radiation, laser chirp, single-color or two-color pump, spectral width of driving radiation, pulse duration, and so on), which was confirmed during experimental and theoretical studies of the HHG in a Sn plasma [53, 68–70]. In the meantime, a further search of superior properties of this plasma, together with consideration of the role of different plasma species (neutrals, singly and doubly-charged ions) in the optimization of efficient harmonic generation, can improve the understanding of the role of various ionic transitions in the efficiency of this nonlinear optical process. Application of high pulse repetition rate lasers generating broadband ultrashort pulses can also enhance the output power of generating single harmonics from tin plasma.

Below, we discuss the theoretical studies of the photoabsorption spectra of different ions in the tin plasma [71] and analyze the calculations of the harmonic output for odd and even harmonics at variable experimental conditions based on the approach developed in [53].

In the published data on the Sn II transitions in the studied spectral region [72], one can find that the observed enhancement of the 16th harmonic ($E = 25.43$ eV) and the 17th harmonic ($E = 27.02$ eV) of the chirp-free 780 nm radiation can be attributed to the transitions $4d^{10}5s^25p\ ^2P_{3/2} \rightarrow 4d^95s^25p^2$. The frequencies of these transitions, some of which possess reasonably large oscillator strengths, lie in the photon energy range of 24.9–27.3 eV. One can attribute the observed enhancement of the 15th harmonic (generated by the negatively chirped pulse) to its approach toward the range of the $4d^{10}5s^25p\ ^2P_{3/2} \rightarrow 4d^95s^25p^2$ transitions of the Sn II ion. However, Sn II transitions cannot explain adequately the enhancement of the 17th harmonic when the chirp variations cause a blue shift and its subsequent generation at a wavelength of 45.1 nm ($E = 27.49$ eV). Therefore, one has to also consider the Sn III ion as a potential source of enhancement of this harmonic.

In order to simulate the resonant HHG enhancement in Sn II and Sn III ions, one has to calculate the properties of the autoionizing (AI) states in these ions. In [72], Duffy et al. recorded

4d–5p photoabsorption spectra of Sn II and Sn III in the 23–33 eV range. Transitions from the $4d^{10}5s^{2}5p$ ground state of Sn II and from the $4d^{10}5s^{2}$ ground and $4d^{10}5s5p$ excited states of Sn III were observed and successfully identified with the aid of multiconfiguration Hartree–Fock calculations. These calculations were repeated in the study under discussion to replicate these spectra and provide estimates for the autoionizing widths of the dominant transitions.

For the Sn II (I_{p} = 14.63 eV) spectrum configuration, the interaction calculations were performed in the Russell–Saunders coupling scheme with the HXR mode (Hartree plus exchange plus relativistic corrections) of the Cowan code [73] for transitions from $4d^{10}5s^{2}5p \rightarrow 4d^{9}5s^{2}5p$, np, mf ($5 \leq n \leq 13$, $4 \leq m \leq 13$). The Slater parameters F^{k}, G^{k}, and the configuration interaction parameter R^{k} were fixed at 85% of the ab initio values while the spin-orbit parameter was left unchanged. Good agreement was achieved between the calculated transition energies and *gf* values and the previous work [72], with maximum discrepancies of 0.12 eV and 0.09, respectively, which are attributed to the increased basis set and scaling factors in the discussed study. Calculations were performed to determine the autoionizing decay widths of the excited 4d states, which are allowed to decay by the following processes:

$$4d^{9}5s^{2}5pnp, mf \rightarrow 4d^{10}5s^{2} + \varepsilon l \quad (l = 0, 2, 4) \tag{2.18}$$

$$\rightarrow 4d^{10}5s5p + \varepsilon' l \quad (l = 1, 3) \tag{2.19}$$

$$\rightarrow 4d^{10}5p^{2} + \varepsilon'' l \quad (l = 2). \tag{2.20}$$

The values of ε, ε', and ε'' are the differences in the configuration average energies of the excited $4d^{-1}$ configuration and each final Sn III ionic configuration. Synthetic spectra were constructed by assuming a Lorentzian line profile, σ (E) = 109.7 $f_k\, \Gamma_k/\ (2\pi\ [(E_k - E)^2 + \Gamma_k^2/\ 4])$, where E_k and Γ_k are the energy and autoionization decay width of the transition in eV and f_k is the oscillator strength. The synthetic spectrum was then convolved with a Gaussian instrumental function of width 0.030 eV, which is presented in Fig. 2.4.

For the Sn III spectra, calculations were performed in *jj* coupling for transitions from both the $4d^{10}5s^{2}$ ground and the $4d^{10}5s5p$ excited configuration, which has a configuration average energy

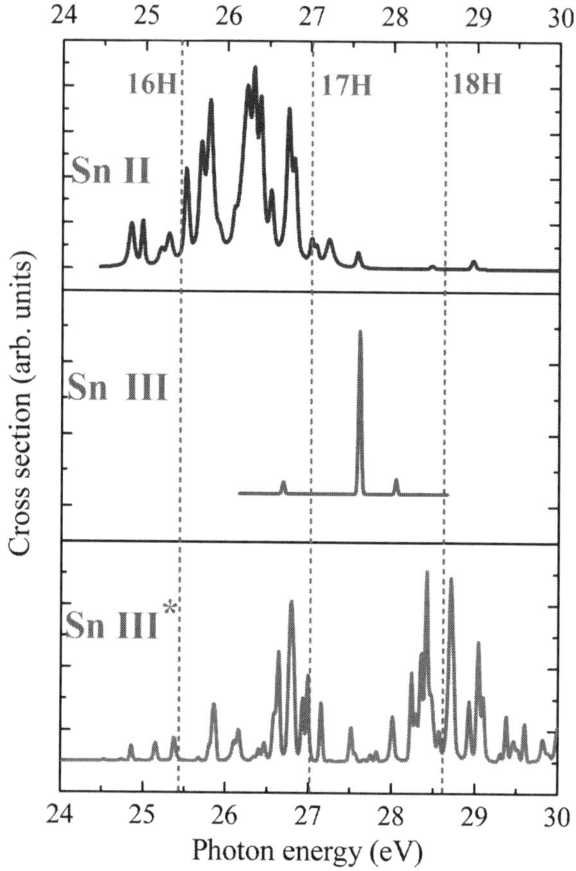

Figure 2.4 The photoabsorption cross-section spectra of Sn II $4d^{10}5s^25p \rightarrow 4d^95s^25p^2$ (top), Sn III $4d^{10}5s^2 \rightarrow 4d^95s^25p$ (middle), and Sn III* excited state $4d^{10}5s5p \rightarrow 4d^95s5p^2$ (bottom) convolved with a Gaussian instrumental function of width 30 meV. Dashed lines show 780 nm radiation harmonic frequencies. Reproduced from Ref. 71 with permission from American Physical Society. Copyright 2012.

of 6.87 eV. In these calculations, the Slater integrals were scaled to 80% except the spin-orbit integral, which was left unchanged. The $4d^95s^25p$ configuration (the configuration average energy of 26.91 eV) does not autoionize because it lies below the ionization limit of Sn IV (30.50 eV [75]). The excited configuration $4d^95s5p^2$ has a configuration average energy of 35.22 eV and therefore once populated from the $4d^{10}5s5p$ excited state can autoionize through the following processes:

$$4d^9 5s5p^2 \rightarrow 4d^{10}5s + \varepsilon l \quad (l = 0, 2, 4), \tag{2.21}$$

where the values of ε are the differences in the configuration average energies of the excited $4d^{-1}$ configuration and each final Sn IV ionic configuration. The resulting synthetic spectra, which were convolved with a Gaussian instrumental function of width 0.030 eV, are also presented in Fig. 2.4. The transition properties required for the resonant HHG simulation are listed in Table 2.1. The experimentally observed enhancement of the blue shifted 17th harmonic (at $\lambda = 45.1$ nm) [71] can be attributed to the Sn III transitions $4d^{10}5s^2 \rightarrow 4d^9 5s^2 5p$ ($E = 27.6$ eV, $\lambda = 44.92$ nm). One can note that autoionizing states are not involved in this transition; thus, this enhancement cannot be explained via a four-step model [53]. The enhancement can be due to better phase-matching conditions near the resonance [48], and/or due to the mechanism of the single-atom response enhancement. Understanding of the origin of this enhancement requires further studies.

Resonant HHG simulation is based on the numerical solution of the 3D TDSE for a model ion in the external laser field. One can use a single-active electron approximation, reproducing the interaction with other electrons and with the nucleus with a model potential, as was done in Refs. [53, 56]. This method is applicable, in particular, for the description of the resonances in the Sn II and Sn III ions.

Table 2.1 The autoionizing properties of some transitions of the tin ions

Ion	Transition	E_{calc} (eV)	gf^e	Γ (meV)
Sn II	$4d^{10}5s^2 5p\ ^2P_{3/2} \rightarrow 4d^9\ 5s^2 5p^2\ ^2D_{5/2}$	26.22[a]	1.43	160.0
Sn III	$4d^{10}5s^2\ ^1S_0 \rightarrow 4d^9 5s^2 5p\ ^1P_1$	27.6[b]	0.870	-
Sn III*	$4d^{10}5s5p\ (1/2,3/2)_2 \rightarrow 4d^9 5s5p^2\ (5/2,1/2)_3^d$	28.48[c]	0.84	47.20

Source: Reproduced from Ref. 71 with permission from American Physical Society. Copyright 2012.

[a,c]Calculated energies were shifted by 0.46 eV and −0.65 eV.

[b]Calculated energies were shifted by 0.16 eV.

[d]The *jj* coupling is denoted $(J_{core}, J_{nl})_J$ where the subscript J refers to the total angular momentum of the level.

[e]The *gf* value is the product of the oscillator strength *f* of a transition and the statistical weight *g* of the lower level.

The following form of the model potential of the parent ion was used (atomic units are used throughout):

$$V(r) = -\frac{Q+1}{\sqrt{a_0^2 + r^2}} + a_1 \exp\left[-\left(\frac{r-a_2}{a_3}\right)^2\right], \tag{2.22}$$

where Q is the charge state of the generating ion (1 for Sn II and 2 for Sn III), and a_0, a_1, a_2, a_3 are the fitting parameters.

These parameters are chosen to reproduce the properties of certain transitions in Sn II and Sn III* ions. Namely, Sn II has a transition $4d^{10}5s^25p\,^2P_{3/2} \rightarrow 4d^95s^25p^2\,(^1D)\,^2D_{5/2}$ with a frequency ($E = 26.22$ eV) close to the 16th and 17th harmonic frequencies of the 780 nm chirp-free driving radiation and with an oscillator strength essentially exceeding the other transitions in this spectral region. Thus, one can neglect the other AI states in this ion and choose the potential parameters to reproduce this AI state energy, the ground state energy, the AI state width, and the oscillator strength of the transition. For Sn III*, the transition with frequency 28.48 eV has the largest oscillator strength in this spectral region. Simulating HHG with Sn III*, one can choose the model potential parameters to reproduce the properties of this transition. The method of the TDSE solution is described in [75]. In Fig. 2.5a, the spectrum of the microscopic response calculated for the Sn II ion is presented. One can see the pronounced enhancement of the 17th harmonic, in agreement with experimental results reported in [71].

To study the frequency range in which the harmonic is enhanced, the HHG in Sn II was calculated using slightly different fundamental frequencies. The harmonic enhancement as a function of its frequency is presented in Fig. 2.5b. One can see that maximal enhancement is achieved for a frequency different to the transition frequency in the absence of the laser, in agreement with experiments [69]. This difference can be attributed to the Stark shift of the AI and ground states in the laser field. The width of the spectral range where the harmonic is enhanced is about 0.7 eV (FWHM). This is higher than the AI width calculated in the absence of the laser field (0.160 eV, see Table 2.1) used in this model. This peculiarity can be explained with the broadening of the AI state due to its photoionization by the laser field.

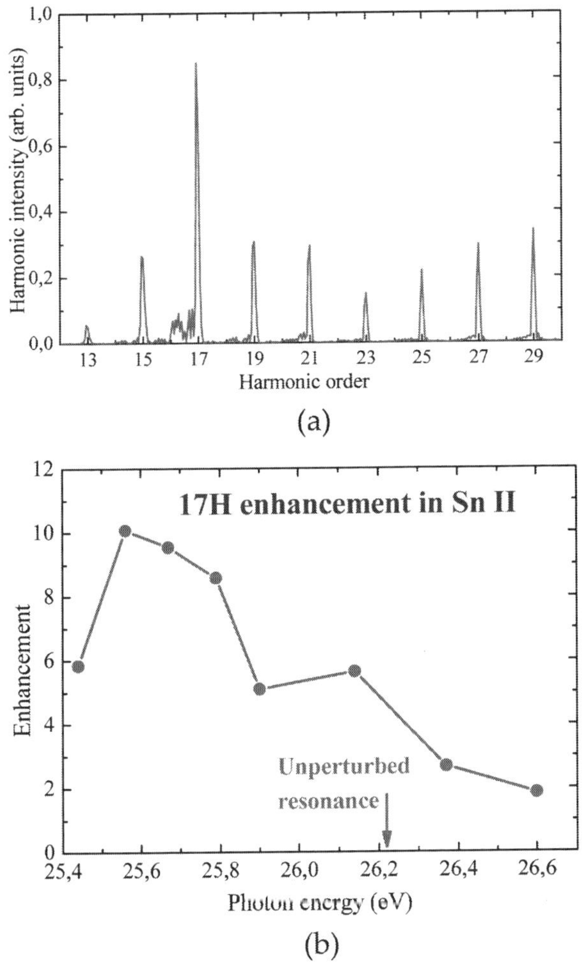

Figure 2.5 (a) Harmonic spectrum calculated for Sn II. The laser intensity is 10^{15}W/cm^2. (b) The calculated resonant 17th harmonic enhancement as a function of the harmonic photon energy. The arrow shows the frequency of the transition in the absence of the laser field. Reproduced from Ref. 71 with permission from American Physical Society. Copyright 2012.

2.5 Calculations of Single Harmonic Generation from Manganese Plasma

In this section, we present the results of numerical simulations of the harmonic generation in manganese plasma within a 1D model [76]. It was assumed that the main contribution to the resonant peak in the spectrum observed in experiment [76] comes from Mn^+ ions. Note that the ionization potential of Mn^{2+} ions (33.7 eV) is more than twice higher than the ionization potential of Mn^+ ions (15.6 eV). The time-dependent Schrödinger equation was solved by means of the split-operator method [77]. The Mn^+ target is modeled using a potential supporting a metastable state by a potential barrier [53, 56]. The shape of the potential is (see Fig. 2.6a)

$$\begin{aligned} V(x) = &-a + a/\{1+\exp[(x+b)/c]\}) + a/\{1+\exp[(-x+b)/c]\} \\ &+ d/(e+x^2)/\{1+\exp[(x+b)/c]\} \\ &+ d/(e+x^2)/\{1+\exp[(-x+b)/c]\}, \end{aligned} \tag{2.23}$$

where a, b, c, d, and e are the parameters. They were chosen to be 1.672, 1.16, 0.216, 8.95, and 0.63, respectively, so that the width of the resonance and the energy gap between the ground and the resonant states resemble the experimental data [78]. The metastable state of this model potential is at 51.8 eV above the ground state. The laser field is $E(t) = E_0 f(t) \cos(\omega_0 t + \phi)$, where $f(t)$ is the pulse envelope, ϕ denotes the carrier envelope phase (CEP), and ω_0 is the laser frequency corresponding to the central wavelength $\lambda = 760$ nm. The laser intensity is $I_0 = 4 \times 10^{14}$ W/cm^2. A CEP of $\phi = 0$ means that the maximum of the envelope corresponds to a maximum of $\sin(\omega_0 t)$.

HHG spectra were calculated for pulse shapes with different lengths and for different values of ϕ (Fig. 2.6b,c). A $\sin^2$ envelope with a total length of four full cycles was used to model the 3.5 fs pulse, while an envelope with four cycles $\sin^2$ switch-on/off, 13 cycles of constant intensity, and 21 cycles total duration was used to model the 40 fs case. The long pulse led to an HHG spectrum that shows well-defined peaks at the odd harmonic orders and that is weakly dependent on the CEP (Fig. 2.6b). Figure 2.6c shows the dependence of the harmonic spectrum on the CEP in the case of the short, few-cycle pulse. In all cases, the resonance dominated the

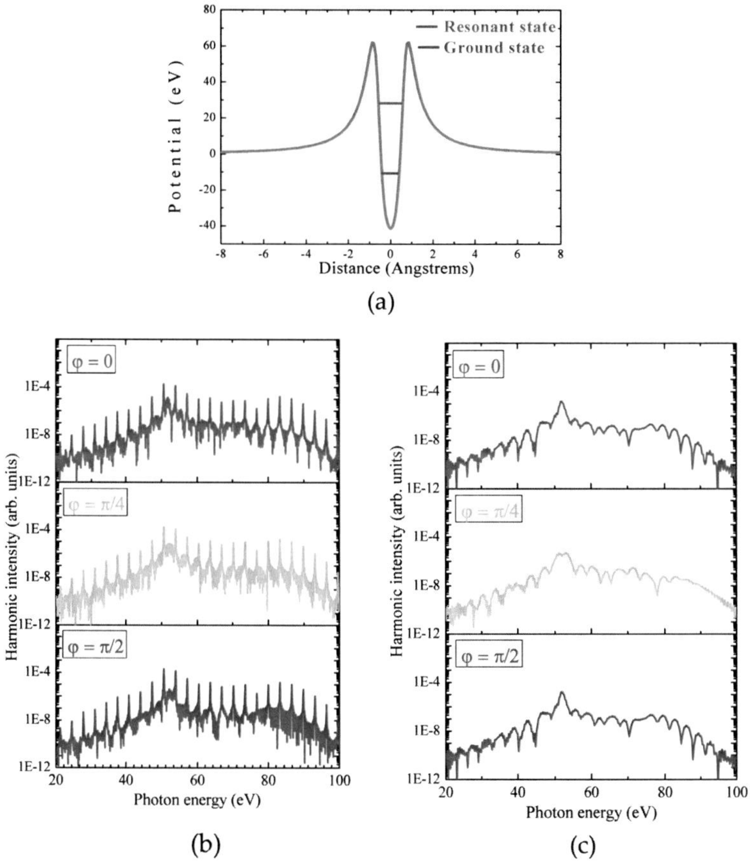

Figure 2.6 (a) Potential used for the numerical simulations. (b,c) Calculated HHG spectra using (b) a long (40 fs) pulse and (c) a few-cycle pulses at different values of the CEP ($\phi = 0$, $\pi/4$, and $\pi/2$). Reproduced from Ref. 76 with permission from Optical Society of America.

spectrum. The most intense emission occurs around 51 eV, where the metastable state is located. Although some difference between harmonic spectra for $\phi = 0$, $\pi/4$, and $\pi/2$ is found, we note that the CEP dependence is strongest for the spectrum outside the region of the resonance. For random CEP, the substructure of the spectrum will average out as confirmed by numerical averaging over 20 values of the CEP in the range from 0 to π. The resonance peak itself

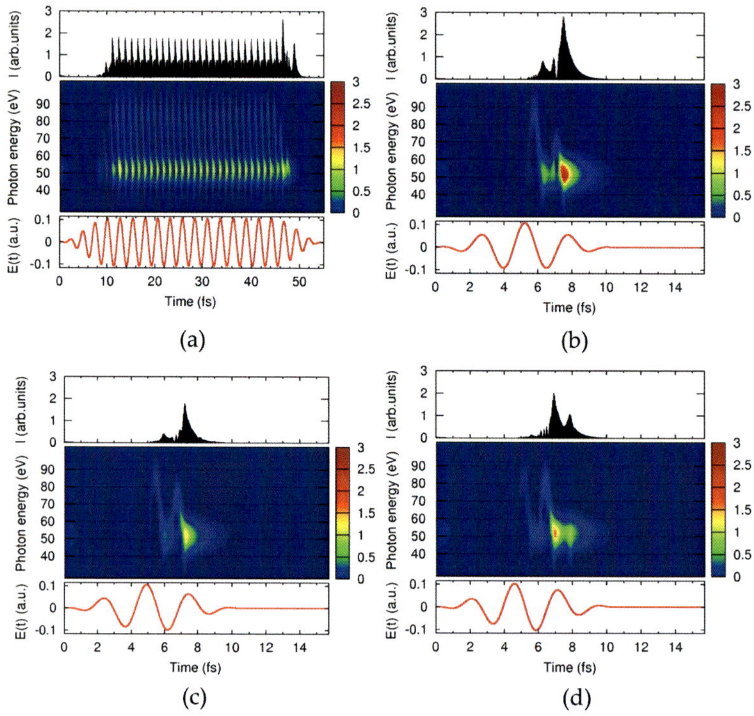

Figure 2.7 Calculated results for the HHG driven by (a) a long (40 fs) pulse with CEP $\phi = \pi/4$ and (b–d) a few-cycle pulses with CEPs of (b) $\phi = 0$, (c) $\phi = \pi/4$, and (d) $\phi = \pi/2$. The top panels show the HHG temporal intensity profile obtained as the square of the time-dependent dipole acceleration after high pass filtering above 32.7 eV. The middle panels show the time-frequency diagrams. The curves in the bottom panels show the time dependence of the electric field of the driving laser pulse. Reproduced from Ref. 76 with permission from Optical Society of America.

depends less on the CEP. The case of $\phi = \pi/4$ appears to be special, as a dip due to trajectory interference seems to coincide with the resonance peak.

In order to investigate temporal characteristics of the harmonic emission in numerical simulations, a Gabor transformation was performed [79]:

$$G(\omega, t) = (2\pi\sigma^2)^{-1/2} \int d\tau a(\tau) \exp(i\omega\tau) \exp[(t-\tau)^2/(2\sigma^2)], \tag{2.24}$$

where $a(\tau)$ is the dipole acceleration from the simulation and σ is a parameter taken to be $\sigma = 1/(3\omega_0)$. The modulus squared, $|G(\omega, t)|^2$, is the time-frequency distribution. The temporal intensity profile of the XUV emission is calculated as the square of the time-dependent dipole acceleration after filtering out the photon energies below 1.2 atomic units (corresponding to 32.7 eV). In fact, the emission profile is not affected much by the filtering, as the spectrum is strongly dominated by the resonance. The results are shown in Fig. 2.7. Comparing the short and long-pulse regimes, one can notice that, whereas in Figs. 2.7b–2.9d, emission of the resonance occurs at the end of the few-cycle pulse, Fig. 2.7a shows that the resonance is repopulated and decaying each half cycle of the multi-cycle pulse.

For most CEPs, the emission can be viewed as an isolated sub-fs XUV pulse if the pulse length is defined, in the usual way, as the full width at half maximum. This main emission burst is either preceded or followed by a small side peak. Similar emission profiles are found for three values of the CEP (0, $\pi/4$, and $\pi/2$, Figs. 2.7b–2.7d). The time of maximum emission varies in a range of less than 1 fs with CEP. These calculations show that one can usually get sub-femtosecond XUV pulses or at least nearly 1 fs XUV pulses, for different values of CEP. This is in sharp contrast with the usual strong CEP dependence of isolated attosecond pulse generation [80–82]. This suggests that resonance-induced HHG driven by few-cycle pulses provides a route to isolated XUV attosecond pulse generation with reduced requirements for CEP stabilization.

References

1. T. Siegel, *Imaging of molecular structure and dynamics using laser driven electron recollisions*, PhD Thesis, Imperial College London (2010).
2. J. F. Ward, G. H. C. New, *Phys. Rev.* 185, 57 (1969).
3. B. W. Shore, P. L. Knight, *J. Phys. B: At. Mol. Opt. Phys.* 20, 413 (1987).
4. A. McPherson, G. Ginson, H. Jara, N. Johann, I. A. McIntyre, K. Boyer, C. K. Rhodes, *J. Opt. Soc. Am. B* 4, 595 (1987).
5. M. Ferray, A. L'Huillier, X. F. Li, L. A. Lompré, G. Mainfray, G. Manus, *J. Phys. B: At. Mol. Opt. Phys.* 21, L31 (1988).

6. Z. H. Chang, A. Rundquist, H. W. Wang, M. M. Murnane, H. C. Kapteyn, *Phys. Rev. Lett.* 79, 2967 (1997).
7. M. Schnurer, C. Spielmann, P. Wobrauschek, C. Streli, N. H. Burnett, C. Kan, K. Ferencz, R. Koppitsch, Z. Cheng, T. Brabec, F. Krausz, *Phys. Rev. Lett.* 80, 3236 (1998).
8. J. Diels, W. Rudolph, *Ultrashort laser pulse phenomena: Fundamentals, techniques, and applications on a femtosecond time scale*, Academic Press, London (1996).
9. P. B. Corkum, *Phys. Rev. Lett.* 71, 1994 (1993).
10. M. Lewenstein, Ph. Balcou, M. Yu. Ivanov, A. L'Huillier, P. Corkum, *Phys. Rev. A* 49, 2117 (1994).
11. R. A. Ganeev, *Laser Phys. Lett.* 9, 175 (2012).
12. K. J. Schafer, B. Yang, L. F. Dimauro, K. C. Kulander, *Phys. Rev. Lett.* 70, 1599 (1993).
13. G. Lagmago Kamta, A. D. Bandrauk, *Phys. Rev. A* 71, 053407 (2005).
14. C. F. D. Faria, *Phys. Rev. A*, 76, 043407 (2007).
15. A. D. Shiner, C. Trallero-Herrero, N. Kajumba, H.-C. Bandulet, D. Comtois, F. Légaré, M. Giguère, J-C. Kieffer, P. B. Corkum, D. M. Villeneuve, *Phys. Rev. Lett.* 103, 073902 (2009).
16. R. A. Ganeev, L. B. Elouga Bom, J. Abdul-Hadi, M. C. H. Wong, J. P. Brichta, V. R. Bhardwaj, T. Ozaki, *Phys. Rev. Lett.* 102, 013903 (2009).
17. M. F. Ciappina, A. Becker, A. Jaron-Becker, *Phys. Rev. A* 76, 063406 (2007).
18. M. Ruggenthaler, S. V. Popruzhenko, D. Bauer, *Phys. Rev. A* 78, 033413 (2008).
19. G. P. Zhang, *Phys. Rev. Lett.* 95, 047401 (2005).
20. H.-D. Meyer, U. Manthe, L. S. Cederbaum, *Chem. Phys. Lett.* 165, 73 (1990).
21. P. V. Redkin, R. A. Ganeev, *Phys. Rev. A* 81, 063825 (2010).
22. V. R. Bhardwaj, P. B. Corkum, D. M. Rayner, *Phys. Rev. Lett.* 93, 043001 (2004).
23. D. Cricchio, P. P. Corso, E. Fiordilino, G. Orlando, F. Persico, *J. Phys. B: At. Mol. Opt. Phys.* 42, 085404 (2009).
24. R. A. Ganeev, C. Hutchison, T. Witting, F. Frank, S. Weber, W. A. Okell, E. Fiordilino, D. Cricchio, F. Persico, A. Zaïr, J. W. G. Tisch, J. P. Marangos, *JOSA B* 30, 7 (2012).
25. R. A. Ganeev, L. B. Elouga Bom, M. C. H. Wong, J.-P. Brichta, V. R. Bhardwaj, P. V. Redkin, T. Ozaki, *Phys. Rev. A* 80, 043808 (2009).

26. R. A. Ganeev, H. Singhal, P. A. Naik, J. A. Chakera, A. K. Srivastava, T. S. Dhami, M. P. Joshi, P. D. Gupta, *J. Appl. Phys.* 106, 103103 (2009).
27. N. Moiseyev, M. Lein, *J. Phys. Chem. A* 107, 7181 (2003).
28. Z.-Y. Zhou, J.-M. Yuan, *Chin. Phys. Lett.* 24, 683 (2007).
29. T. Millack, A. Maquet, *J. Mod. Opt.* 40, 2161 (1993).
30. F. I. Gauthey, C. H. Keitel, P. L. Knight, A. Maquet, *Phys. Rev. A* 52, 525 (1995).
31. W. Chu, Y. Xie, S. Duan, N. Yang, W. Zhang, J.-L. Zhu, X.-G. Zhao, *Phys. Rev. B* 82, 125301 (2010).
32. Z.-Y. Zhou, J.-M. Yuan, *Phys. Rev. A* 77, 063411 (2008).
33. V. Kapoor, D. Bauer, *Phys. Rev. A* 85, 023407 (2012).
34. A. Di Piazza, E. Fiordilino, *Phys. Rev. A* 64, 013802 (2001).
35. A. D. Bandrauk, S. Chelkowski, H. S. Nguyen, *J. Mol. Struc.* 735–736, 203 (2005).
36. U. Fano, *Phys. Rev.* 124, 1866 (1961).
37. G. Raşeev, B. Leyh, H. Lefebvre-Brion, Z. *Phys. D: At. Mol. Clusters* 2, 319 (1986).
38. F. Keller, H. Lefebvre-Brion, Z. *Phys. D: At. Mol. Clusters* 4, 15 (1986).
39. M. Y. Amusia, J.-P. Connerade, *Rep. Prog. Phys.* 63, 41 (2000).
40. J. F. Reintjes, *Nonlinear Optical Parametric Processes in Liquids and Gases, Academic Press*, New York (1984).
41. C. Figueira de Morisson Faria, R. Kopold, W. Becker, J. M. Rost, *Phys. Rev. A* 65, 023404 (2002).
42. R. Taieb, V. Veniard, J. Wassaf, A. Maquet, *Phys. Rev. A* 68, 033403 (2003).
43. M. B. Gaarde, K. J. Schafer, *Phys. Rev. A* 64, 013820 (2001).
44. E. S. Toma, P. Antoine, A. de Bohan, H. G. Muller, *J. Phys. B: At. Mol. Opt. Phys.* 32, 5843 (1999).
45. Z. Zeng, R. Li, Y. Cheng, W. Yu, Z. Xu, *Phys. Scripta* 66, 321 (2002).
46. R. Bartels, S. Backus, E. Zeek, L. Misoguti, G. Vdovin, I. P. Christov, M. M. Murnane, H. C. Kapteyn, *Nature* 406, 164 (2000).
47. L. Plaja, L. Roso, *J. Modern Opt.* 40, 793 (1993).
48. R. A. Ganeev, *Open Spectrosc. J.* 3, 1 (2009).
49. R. A. Ganeev, P. A. Naik, H. Singhal, J. A. Chakera, P. D. Gupta, *Opt. Lett.* 32, 65 (2007).
50. R. A. Ganeev, P. A. Naik, J. A. Chakera, H. Singhal, N. C. Pramanik, P. A. Abraham, N. Rani Panicker, M. Kumar, P. D. Gupta, *J. Opt. Soc. Am. B* 28, 360 (2011).

51. D. B. Milošević, *J. Phys. B: At. Mol. Opt. Phys.* 40, 3367 (2007).
52. I. A. Kulagin, T. Usmanov, *Opt. Lett.* 34, 2616 (2009).
53. V. Strelkov, *Phys. Rev. Lett.* 104, 123901 (2010).
54. D. B. Milošević, *Phys. Rev. A* 81, 023802 (2010).
55. M. V. Frolov, N. L. Manakov, A. F. Starace, *Phys. Rev. A* 82, 023424 (2010).
56. M. Tudorovskaya, M. Lein, *Phys. Rev. A* 84, 013430 (2011).
57. M. V. Frolov, N. L. Manakov, T. S. Sarantseva, M. Yu. Emelin, M. Yu. Ryabikin, A. F. Starace, *Phys. Rev. Lett.* 102, 243901 (2009).
58. S. Minemoto, T. Umegaki, Y. Oguchi, T. Morishita, A. T. Le, S. Watanabe, H. Sakai, *Phys. Rev. A* 78, 061402 (2008).
59. H. J. Wörner, H. Niikura, J. B. Bertrand, P. B. Corkum, D. M. Villeneuve, *Phys. Rev. Lett.* 102, 103901 (2009).
60. L. B. Elouga Bom, Y. Petrot, V. R. Bhardwaj, and T. Ozaki, *Opt. Express* 19, 3077 (2011).
61. P. Agostini, L. F. DiMauro, *Rep. Prog. Phys.* 67, 813 (2004).
62. F. Krausz, M. Ivanov, *Rev. Mod. Phys.* 81, 163 (2009).
63. P. V. Redkin, M. K. Kodirov, R. A. Ganeev, *J. Opt. Soc. Am. B* 18, 165 (2011).
64. E. Runge, E. K. U. Gross, *Phys. Rev. Lett.* 52, 997 (1984).
65. M. A. L. Marques, A. Castro, G. F. Bertsch, A. Rubio, *Comput. Phys. Commun.* 151, 60 (2003).
66. A. Castro, H. Appel, M. Oliveira, C. A. Rozzi, X. Andrade, F. Lorenzen, M. A. L. Marques, E. K. U. Gross, A. Rubio, *Phys. Status Solidi B* 243, 2465 (2006).
67. M. A. L. Marques, C. A. Ullrich, F. Nogueria, K. Burke, E. K. U. Gross, *Time-dependent Density Functional Theory*, Springer, Heidelberg (2003).
68. M. Suzuki, M. Baba, R. Ganeev, H. Kuroda, T. Ozaki, *Opt. Lett.* 31, 3306 (2006).
69. R. A. Ganeev, L. B. Elouga Bom, J.-C. Kieffer, T. Ozaki, *Phys. Rev. A* 75, 063806 (2007).
70. R. A. Ganeev, J. A. Chakera, P. A. Naik, H. Singhal, R. A. Khan, P. D. Gupta, *J. Opt. Soc. Am. B* 28, 1055 (2011).
71. R. A. Ganeev, V. V. Strelkov, C. Hutchison, A. Zaïr, D. Kilbane, M. A. Khokhlova, J. P. Marangos, *Phys. Rev. A* 85, 023832 (2012).
72. G. Duffy, P. van Kampen, P. Dunne, *J. Phys. B: At. Mol. Opt. Phys.* 34, 3171 (2001).
73. R. D. Cowan, *The Theory of Atomic Structure and Spectra*, University of California Press, Berkeley (1981).

74. D. R. Lide, *CRC Handbook of Chemistry and Physics* (73rd ed.), Academic Press, New York (1992).
75. V. Strelkov, A. Sterjantov, N. Shubin, V. Platonenko, *J. Phys. B: At. Mol. Opt. Phys.* 39, 577 (2006).
76. R. A. Ganeev, T. Witting, C. Hutchison, F. Frank, M. Tudorovskaya, M. Lein, W. A. Okell, A. Zaïr, J. P. Marangos, J. W. G. Tisch, *Opt. Express* 20, 25239 (2012).
77. M. D. Feit, J. A. Fleck Jr, A. Steiger, *J. Comp. Phys.* 47, 412 (1982).
78. H. Kjeldsen, F. Folkmann, B. Kristensen, J. B. West, J. E. Hansen, *J. Phys. B: At. Mol. Opt. Phys.* 37, 1321 (2004).
79. D. Gabor, *J. Inst. Electr. Eng.* 93, 429 (1946).
80. C. Altucci, J. W. G. Tisch, R. Velotta, *J. Mod. Opt.* 58, 1585 (2011).
81. L. E. Chipperfield, L. N. Gaier, P. L. Knight, J. P. Marangos, J. W. G. Tisch, *J. Mod. Opt.* 52, 243 (2005).
82. E. Goulielmakis, M. Schultze, M. Hofstetter, V. S. Yakovlev, J. Gagnon, M. Uiberacker, A. L. Aquila, E. M. Gullikson, D. T. Attwood, R. Kienberger, F. Krausz, U. Kleineberg, *Science* 320, 1614 (2008).

Chapter 3

Low-Order Harmonic Generation in Plasma Plumes Using Nanosecond and Picosecond Driving Pulses

Initially, low-order harmonic generation using relatively long pulses was aimed to define the conditions of efficient third harmonic (3H) generation in plasma plumes [1]. The advantages of plasma harmonics could largely be realized with the use of a low-excited and weakly ionized plasma, because the limiting processes governing the dynamics of laser frequency conversion would play a minor role in this case. Attention was drawn to this feature early in the study of third harmonic generation (THG) in a weakly ionized plasma. An analysis of low-order (third [2] and fifth [3]) harmonic generation in such plasmas produced by a laser prepulse on the surface of a solid target allowed formulating several recommendations with regard to further advancement toward shorter wavelengths. The 3H conversion efficiency in that case was $\leq 10^{-3}$.

Some stronger 3H conversion efficiency was observed in the spark from the optical breakdown of an atmospheric pressure gas near a metal surface [4]. The breakdown was produced by a nanosecond Nd:YAG laser with 250 mJ pulse energy and pulse duration of 15 ns. The laser that was used for 3H generation had

Plasma Harmonics
Rashid A. Ganeev

ISBN 978-981-4463-30-0 (Hardcover), 978-981-4463-31-7 (eBook)
www.panstanford.com

a pulse duration of 40 ps and energy of 50 mJ. One can note considerably higher concentration of medium in this case (of the order of 5×10^{19} cm^{-3}) than moderate ablation plumes used in plasma harmonic experiments (10^{17}–10^{18} cm^{-3}).

In the following sections, I analyze some recent studies of low-order (third, fourth, and fifth) harmonic generation in laser plasmas using relatively long pulses.

3.1 Low-Order Harmonic Generation in Metal Ablation Plasmas in Nanosecond and Picosecond Regimes

Harmonic generation (HG) of intense laser pulses in isotropic media such as gases and vapors serves to create coherent radiation in the short wavelength range of the spectrum down to the XUV and X-ray regions. Low-order frequency upconversion using laser pulses of nanosecond duration has provided table-top vacuum ultraviolet (VUV) coherent light for spectroscopy and molecular photodissociation studies [5, 6]. On the contrary, HHG requires pulses of higher intensities (typically 10^{14}–10^{15} W/cm^2), which can easily be achieved with femtosecond driving lasers. Although the phenomenon of HG is universal and can be generalized to any material with sufficient nonlinear response, mostly atomic [7, 8], or in some cases, molecular [9, 10] gas jets have been investigated as nonlinear media. In practical applications of coherent short wavelength HG sources, a high conversion efficiency and thus high photon flux of the harmonic radiation are essential.

For understanding the impeding and restricting processes of HG in laser-generated plasmas and identifying the media that lead to highly efficient up-conversion, the analysis of low-order harmonics can be a clearly justified approach, as, for lowest orders, one can clearly define the processes, which hamper further enhancement of conversion efficiency. Examples of this approach are recent investigations of low-order HG in ablation plumes of dielectric and semiconductor solid targets using Q-switched Nd:YAG lasers with pulses of nanosecond duration [11, 12].

The studies in two time domains embracing the nanosecond and picosecond scales of the low-order HG in ablation plasmas of the manganese (Z = 25), copper (Z = 29), and silver (Z = 47), materials are presented in [13], which have already shown efficient nonlinear up conversion of laser wavelength into the UV and XUV ranges [14, 15]. I will discuss below the results of this work showing a comparative analysis of low-order HG using different regimes of target ablation and conversion of laser radiation.

In the first case, two synchronized nanosecond Q-switched Nd:YAG lasers emitting at 1064 nm were used for triggering the ablation event and as fundamental radiation for the HG process. In the second case, a picosecond Nd:YAG laser also operating at 1064 nm served both for ablation and as a source of fundamental radiation for frequency up-conversion.

In this work, account was given of the distinct spatial and temporal distribution of the harmonic signals across the plasmas, and of the differences observed between nanosecond and picosecond regimes. The interest of this work relies on the capability of low-order HG to monitor in situ the composition and dynamics of ablation plasma plumes and to define optimal schemes and plasmas, based in the widely used Nd:YAG lasers, for further development of coherent sources in the VUV near and below 200 nm range, a region where frequency doubling in nonlinear crystals is extremely restricted.

Two setups were employed, corresponding to experiments in the nanosecond and picosecond regimes. For HG with nanosecond pulses [11, 12], ablation plasmas of manganese, copper, and silver were generated in vacuum through normal incidence irradiation with a Q-switched Nd:YAG laser [6 ns full-width at half-maximum (FWHM), 1064 nm] that was operated at 10 Hz using pulse energies of 2–35 mJ. This ablation beam was focused with a 17 cm focal length lens on the surface of the metal target down to a spot of 0.4 mm diameter, generating maximum intensities of 4.5×10^9 W/cm^2, above the ablation thresholds. The vacuum chamber was kept at 10^{-2} mbar background pressure, and the samples were mounted on a rotating holder to avoid cratering. The harmonic driving IR fundamental radiation was delivered by a second Nd:YAG laser (15 ns FWHM, 1064 nm, 10 Hz) propagating parallel to the target surface in such a way that it intersected the ablation plasma at a controllable distance from the surface (distances of 0.6 and 1 mm were employed). Typical pulse energies were in the range of 20–650 mJ. The beam was focused with a 20 cm focal length lens to a spot of 0.040 mm diameter, so that laser intensities in the harmonic generating region reached values of 1.0×10^{11} to 3.5×10^{12} W/cm^2. The delay between this driving pulse and the ablation pulse was controlled electronically in the range from 0 to 10 μs.

Low-order harmonics of the driving laser were produced in the interaction region, and those propagated collinearly with the driving beam. Two highly IR transmissive, UV reflective mirrors were placed at the exit of the vacuum chamber to avoid damage of the detector with the intense IR beam. After the mirrors, a system consisting of a monochromator with a ruled 300 lines/mm grating coupled to a time-gated, intensified charge-coupled device (ICCD) was used for spectral discrimination and detection of the generated harmonics. For measurements, third (TH) and fifth (FH) harmonics were detected in the same spectral window by monitoring, together with the TH at 355 nm, the second order of the FH at 425.6 nm (2FH). Electronically excited species generated in the plasma produced optical emissions that could, under some conditions, be detected simultaneously with the harmonics. Typical acquisitions corresponded to an accumulation of 150 laser shots and a detection gate of 100 ns.

For HG with picosecond pulses, a home-made Nd:YAG laser operating at 1064 nm and delivering 38 ps pulses of up to 33 mJ at a 2 Hz pulse repetition rate was employed. The pulse was divided by a beam splitter into two beams: one was used for ablating the target and the second, as fundamental driving beam, was delayed with respect to the former in the 15–50 ns range using an optical delay line. Typically, a 5 mJ pulse was focused down to a diameter of 0.5 mm by a 17 cm focal length lens on the targets for ablation. The driving beam, with energies per pulse of 0.5–28 mJ, was in turn focused by a 15 cm focal length lens to a focal spot of 30 μm diameter. The intensity of the ablating pulse on the target was 1.3×10^{11} W/cm^2, and that of the probe pulse ranged from 2.0×10^{12} to 1.0×10^{14} W/cm^2. The targets were ablated in vacuum and the up-converted radiation was measured by a fiberoptic spectrometer HR4000.

Harmonics of the third and fifth orders (monitored at 355 nm and in second order of the grating at 425.6 nm, respectively) of the fundamental 15 ns, 1064 nm beam were observed for each of the three metals in a range of ablation and probe laser energies and in a given spatiotemporal plume domain. Spontaneous emissions from excited species in the plume were recorded simultaneously with the signal from harmonics and consisted on the lines corresponding to atomic neutrals and ions. In the case of silver, some broad bands around 409 and 420 nm could also be attributed to AgO molecular emissions. As an example, Fig. 3.1 gives proof of the TH and FH signatures for the case of copper when both

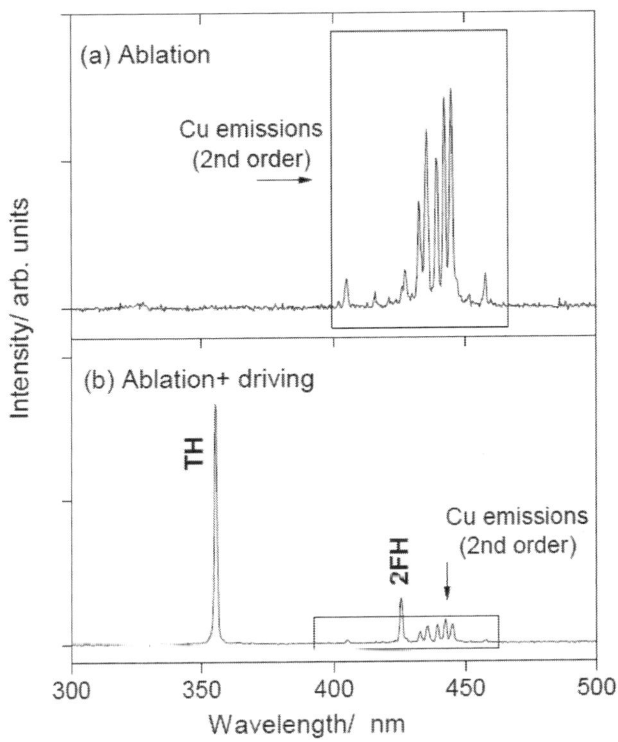

Figure 3.1 Spectra of emissions detected (a) upon ablation of a copper target with 6 ns, 1064 nm pulses of 11 mJ, and in the presence (b) of the fundamental driving beam of the same wavelength propagating at 0.6 mm from the target with a delay of 250 ns. The energy of the driving beam was 560 mJ. The lines corresponding to third harmonic and the second order of the fifth harmonic and of the atomic copper lines are indicated. Reprinted with permission from Ref. 13. Copyright 2012. AIP Publishing LLC.

ablation and fundamental beams were present. The second order of some atomic Cu emissions appears in the 430–440 nm region. Experiments performed with ablation and fundamental laser pulses of picosecond duration emitting at 1064 nm yielded efficient TH for the three metal plasmas, but no signal was detected for the FH, due to the low efficiency of the spectral acquisition system at the corresponding wavelength.

Figure 3.2 shows the behavior of the TH as a function of the pulse energy of the nanosecond ablation laser for each of the three metals. The signal corresponding to the FH (not shown) closely follows that of the TH. In all cases, the onset of harmonic signal coincided with the appearance of a luminous plume, except for manganese, in which harmonic signals were detected at pulse energies somehow lower than those necessary to observe the plume emission. Harmonic signals expectedly increase with ablation energy, as this promotes the ejection of higher amount of material from the target and, thus, the higher local densities of nonlinear emitters (through the expected quadratic dependence with local density [16]). Maximum harmonic yields were observed at ablation pulse energies that increase with the atomic number of the metal, with values of 5, 10, and 12 mJ for Mn, Cu, and Ag respectively. The decrease of the signal at higher pulse energies should be ascribed to the increase of free electron density in the plasma and their detrimental contribution to the phase mismatch. It is observed that when using nanosecond ablation pulses, the Mn plasma can sustain HG in a wider range of ablation energies.

The nonlinear character of the emissions at 355 and 425.6 nm was confirmed by studying the power dependence with respect to the pulse energy of the driving laser beam. Figure 3.3 displays those measurements for 6 ns, 1064 nm laser-generated plasmas of Mn, Cu, and Ag when the driving laser was propagating at a distance of 0.6 mm from the surface. The slopes obtained in log-log plots were compatible with the values of 3 and 5, expected for the behavior of the TH and FH in the perturbative regime. As these measurements were collected for each metal at ablation pulse energies corresponding to maximum up-conversion, that is, 5, 10, and 12 mJ for Mn, Cu, and Ag respectively, they provide an estimation of the relative harmonic signal intensity, where the highest and

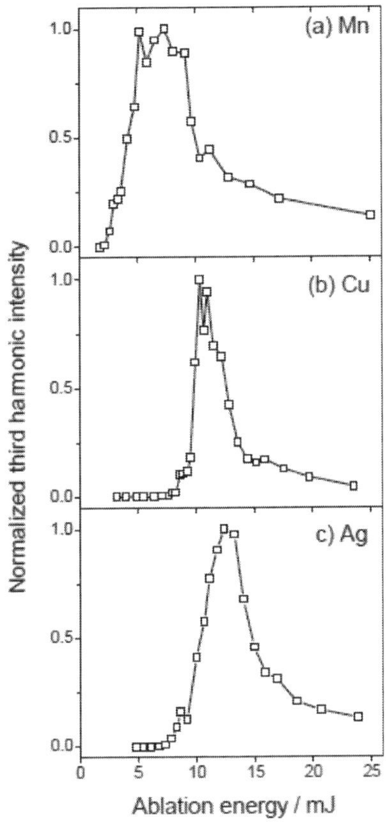

Figure 3.2 Normalized intensity of the third harmonic (355 nm) of the driving laser as a function of the ablation pulse energy for (a) Mn, (b) Cu, and (c) Ag. The conditions were the following: driving laser energy 560 mJ, distance to target 0.6 mm, delay 250 ns. Both ablation and driving laser pulses have durations within nanosecond range. Reprinted with permission from Ref. 13. Copyright 2012. AIP Publishing LLC.

lowest TH yields are obtained for Mn and Ag, respectively. It is interesting to note that for Mn, the FH signal reaches that of the TH at pulse energies of the fundamental beam above 150 mJ, corresponding to an intensity of 7.9×10^{11} W/cm^2. Therefore, by using Mn under the present experimental conditions in the nanosecond domain, it is possible to select the relative harmonic intensity by tuning the pulse energy of the driving beam.

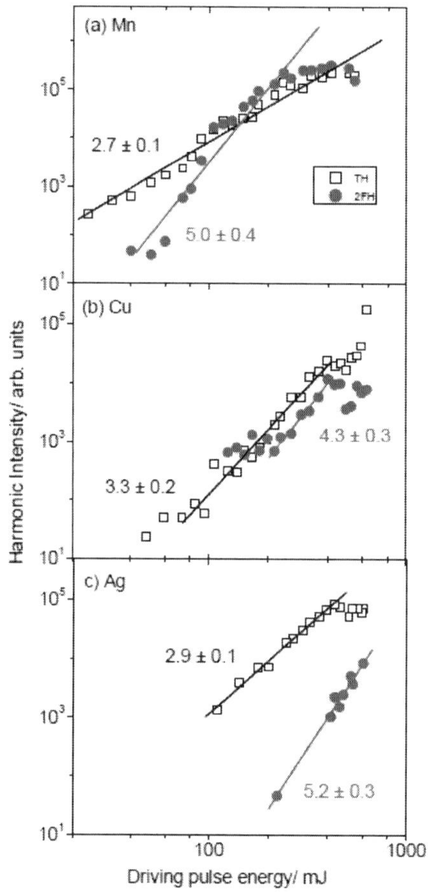

Figure 3.3 Log-log plot of the intensity of the TH (open squares) and of the FH harmonic (solid circles) of the nanosecond driving laser detected at a distance of 0.6 mm from the target as a function of the laser pulse energy, for nanosecond, 1064 nm ablation of metal targets. The conditions were (a) Mn, ablation pulse energy 5 mJ, delay 250 ns; (b) Cu ablation pulse energy 10 mJ, delay 250 ns; and (c) Ag ablation pulse energy 12 mJ, delay 300 ns. As indicated, the slopes are compatible with third-order and fifth-order nonlinear processes. Reprinted with permission from Ref. 13. Copyright 2012. AIP Publishing LLC.

In the picosecond regime, the intensity of TH signal was also monitored as a function of the driving pulse energy and the results are compatible with the expected cubic dependence. Also in this case, Mn displayed the highest conversion efficiency. The intensity of the emitted harmonics was sampled across the temporal-spatial region of the plume. This was done by monitoring on the one hand the dependence with the delay between ablation and driving pulses, and on the other hand, with a set of z-scan measurements, that is, measurements of the harmonic signal as a function of the position z of the focus of the driving laser with respect to the center of the plasma in the direction of propagation, and with the distance of the propagation direction of the driving laser to the target surface (x coordinate).

Figure 3.4a–c shows the emitted TH and FH intensities in the metal plumes in the nanosecond configuration, at 0.6 mm from the surface of the target, as functions of the time delay. The curves obtained were normalized for easier comparison. One temporal component is observed for all three metals. It is observed that the optimal delays, listed in detail in Table 3.1, and the extent of the temporal component are appreciably higher for the lighter element manganese. This is not surprising, as expansion of lighter atoms and ions from the target surface, presumably responsible for the earlier stages of HG [12], reach the region sampled by the fundamental beam earlier as compared with heavier species. Therefore, the highest up-conversion yield is obtained at the optimum phase-matching conditions when the free-electron induced dispersion is overcome by dispersion attributable to atoms, ions, and other

Table 3.1 Position in nanoseconds of the temporal maxima of TH and FH during ablation of metals by nanosecond and picosecond pulses

	Mn		Cu		Ag	
Distance (mm)	0.6	1.0	0.6	1.0	0.6	1.0
TH(ns)	500	600	250	300	400	450
2FH(ns)	500	600	250	300	400	450
Distance (mm)			0.2			
TH(ps)	28		20		30	

Source: Reprinted with permission from Ref. 13. Copyright 2012. AIP Publishing LLC.

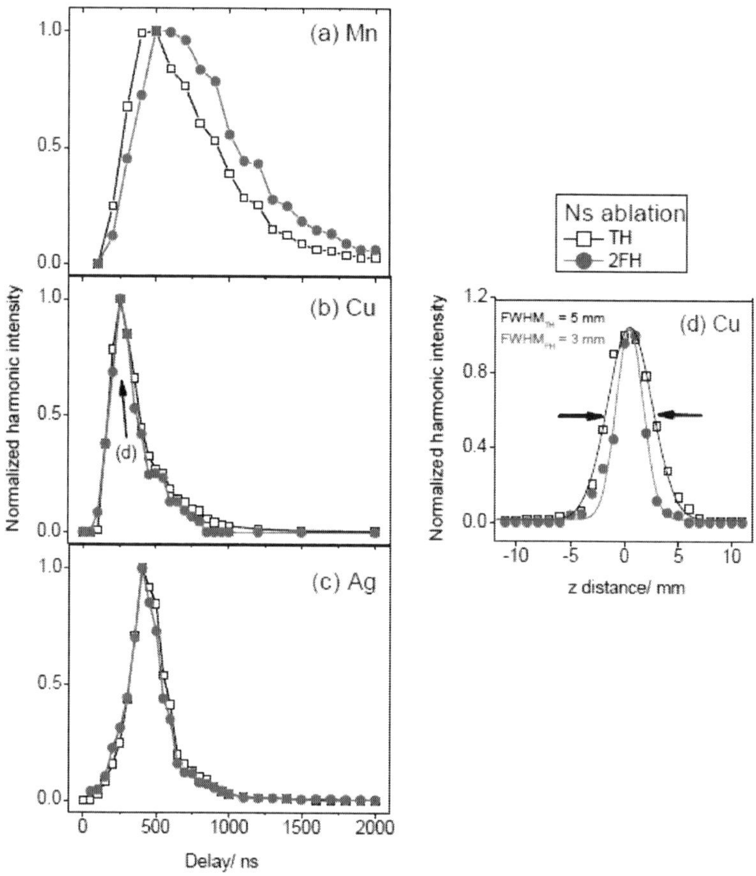

Figure 3.4 Normalized harmonic signals of the third (open squares) and fifth (solid circles) orders of the IR ns driving laser in metal ablation plasmas induced by a 6 ns, 1064 nm ablation laser with pulse energy of 5, 10, and 12 mJ for (a) Mn, (b) Cu, and (c) Ag respectively, as a function of the delay between the ablation event and the arrival of the nanosecond driving laser, and at a distance of 0.6 mm from the surface of the target. The driving laser power was 560 mJ. A *z*-scan plot for copper, in which the TH and FH are measured as the functions of the position of the IR focus (*z*) with respect to the centre of the ablation plasma in the direction of propagation, is represented in (d) for a delay of 250 ns. Experimental points are shown together with Gaussian fits with the widths (FWHM) indicated in the graph. Reprinted with permission from Ref. 13. Copyright 2012. AIP Publishing LLC.

Table 3.2 Full-width half-maxima of harmonic emissions (in mm) as a function of the position of IR driving laser focus with respect to center of ablation plasma in the direction of laser propagation (*z*-scan). The distance of the propagation direction of the driving laser to the target is 0.6 mm for nanosecond pulses and 0.2 mm for picosecond pulses

	Mn	Cu	Ag
TH (ns)	8	5	5
2FH (ns)	5	3	4
TH (ps)	17	14	7

Source: Reprinted with permission from Ref. 13. Copyright 2012. AIP Publishing LLC.

nonlinear species created in the ablation plume (see below). Consistently, further away from the target, at a distance of 1 mm, the optimum delay is temporally displaced as expected (Table 3.1), because the nonlinear emitters ejected from the target take longer time than electrons to reach the plume region that interacts with the driving laser. It is also interesting to note that the relative strength of the third and fifth harmonics is independent of the time delay, indicating that this ratio is not governed by the instantaneous density of nonlinear plume species [17].

With regard to the spatial behavior of the harmonic signal generated by the nanosecond beam across the plume, Fig. 3.4d includes, as an example, the *z*-scan measurements performed for copper under the conditions marked with an arrow in Fig. 3.4b. The *z*-scan presents a single maximum, corresponding to focusing the driving laser at the center of the plasma and this characteristic is also found for Mn and Ag. Table 3.2 summarizes the *z*-scan results obtained by fitting the data with Gaussian functions. A broader structure is observed for the TH beam than for the FH, in agreement with the sharper laser intensity dependence of the latter. It is also found that the FWHM of the harmonic signals for Mn is larger than those corresponding to the other two metals. This indicates that for this lighter metal, the angular distribution of nonlinear species is spatially broader over the plane perpendicular to the normal to the surface, that is, they are emitted over an angularly wider distribution than in the case of Cu and Ag.

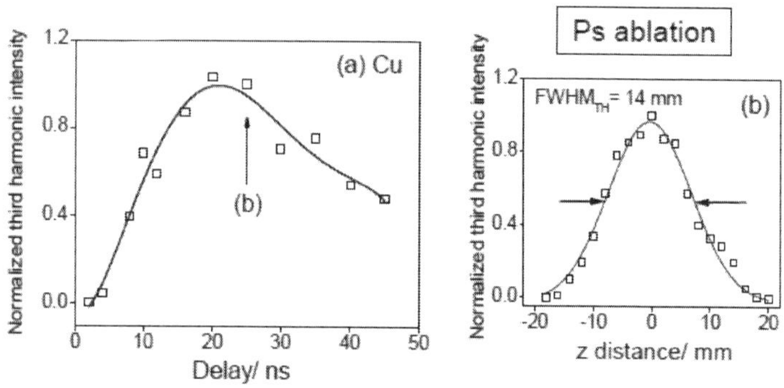

Figure 3.5 (a) Third harmonic of the IR picosecond driving laser in copper ablation plasma induced by a 38 ps, 1064 nm ablation laser with pulse energy of 10 mJ as a function of the delay between the ablation event and the arrival of the driving laser, and at a distance of 0.2 mm from the surface of the target. The driving laser power was 4.5×10^{13} W/cm^2. (b) Z-scan plot for copper, in which the TH is measured as a function of the position of the IR focus (z) with respect to the center of the ablation plasma in the direction of propagation, for a delay of 25 ns. The experimental points are shown together with the Gaussian fit yielding the indicated width (FWHM). Reprinted with permission from Ref. 13. Copyright 2012. AIP Publishing LLC.

The dependence of the TH signal generated in the picosecond domain with the time delay between ablation and probe lasers was also measured for the three studied metals at 0.2 mm from the target. A representative result is shown in Fig. 3.5a for the case of copper. Similar delay dependence is observed for Mn and Ag, and the times of maximum TH signal are listed in Table 3.1. It is important to note that, in striking contrast with the nanosecond results (Fig. 3.4a–c), the harmonic signal decays much faster with delay and vanishes after few tens of nanoseconds. Figure 3.5b displays the z-scan measurements performed on this material. The FWHM of the signal for copper is 14 mm, about a factor of 3 larger than the value obtained in the corresponding nanosecond experiment. Values for Mn and Ag are also listed in Table 3.2. FWHM values derived from z-scan measurements in the nanosecond and picosecond experiments were obtained by probing different plasma regions where efficient HG was observed, that is, at 0.6 mm by the nanosecond driving beam and closer to the target at 0.2 mm by the picosecond beam.

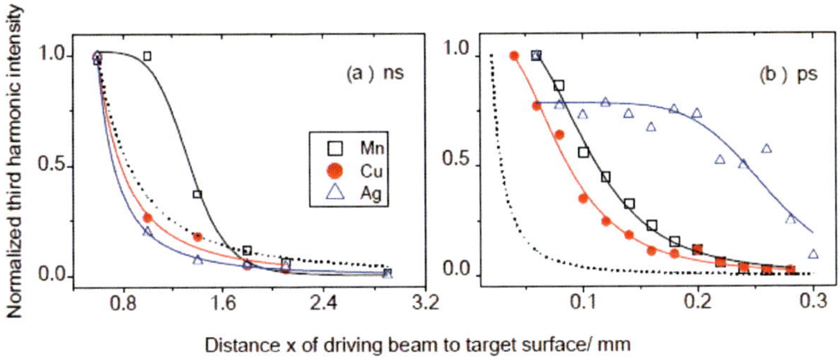

Figure 3.6 Normalized intensity of the TH signal at 355 nm as the position of the fundamental IR beam is displaced from the surface of the metal target [x distance with ablation and driving lasers with pulses of (a) nanosecond and (b) picosecond duration]. For each distance probed, the delay between the ablation beam at 1064 nm and the fundamental beam was set at the value corresponding to the optimum signal. The solid lines are visual guides while the dotted lines represent a x^{-2} decay. Reprinted with permission from Ref. 13. Copyright 2012. AIP Publishing LLC.

The FWHM values (Table 3.2) serve to estimate the extension of nonlinear emitters along the region parallel to the target surface probed by the driving laser.

The measurements of the intensity of the harmonic signals, as a function of the distance to the target x, give additional information on the spatial dimensions of the laser-created nonlinear media (Fig. 3.6). In Fig. 3.6a, the behavior of the TH generated by nanosecond IR laser is plotted for an ablation-driving delay that corresponds to maximum TH. It is observed that the harmonic signal is no longer visible for distances longer than 3 mm away from the target for all three materials. Figure 3.6b shows the corresponding results in the picosecond case, in which it is found that at distances over 0.30 mm from the target, the harmonic signal is negligible. The lesser extent of plume expansion in the direction perpendicular to the target under picosecond ablation, as compared with the nanosecond case, has been observed before in various materials [18].

The dependence of harmonic signal with the distance to the target x is mediated by the corresponding dependence of the product $d^2 L_p^2$, where d is the density and L_p is the length of nonlinear medium [16]. In an isotropic emission of nonlinear species from the target, d scales with x^{-2}. As L_p is expected to be proportional to x, the harmonic signal should be compatible with a x^{-2} dependence. The deviation in the nanosecond experiments from the x^{-2} behavior, indicated by a dotted line in Fig. 3.6a, is appreciable for Mn while Cu and Ag closely follow this dependence. A clear departure of the x^{-2} dependence is observed for all three metals in the picosecond experiments (Fig. 3.6b). It is expected that anisotropic plasma expansion, appearance of clusters in the plume or the presence of plasma electrons, at higher concentration in the picosecond plume due to the higher degree of ionization would contribute to the observed x-dependence of harmonic signal.

The results outlined above allow discussing the identity and spatiotemporal distribution of the species contributing to frequency up-conversion in the metal ablation plasmas. The position of the maxima found on the time delay dependence (Table 3.1, Fig. 3.4) corresponds to velocities of around 1.2×10^5 to 2.4×10^5 cm/s, with the highest velocity for the lower Z-number element, manganese. This range of velocities is typical for ejected species under these nanosecond ablation conditions [12, 18]. For picosecond experiments, the velocity of the nonlinear species derived from the delay dependence plots (Table 3.1, Fig. 3.5) is around 10^6 cm/s.

Assuming a totally statistical ablation process, some insight can be obtained on the nature of nonlinear species contributing to HG in the nanosecond ablation plasma [12]. Under the conditions of thermodynamic equilibrium in the first expansion stages, the same average kinetic energy is expected for all plasma components. Therefore, it is possible to estimate the average size of species contributing to HG in the observed time delay range if one considers that the atomic species are responsible for the early HG stage. For Mn, the temporal delay range of efficient HG, Fig. 3.4a, would correspond to the emission of species up to $(Mn)_{400}$. For copper and silver, the more restricted range of temporal delays for efficient HG, Fig. 3.4b,c, derives in a range of species up to about 60 atoms. For the picosecond laser ablation plasmas, the narrower delay range, of few tens of nanoseconds, for efficient HG provides an indication of the more homogeneous nature of the plasma species, both in terms of velocity and mass. Also in this case, the highest velocity of the ejected species anticipates a lower degree of interactions within the plume species, leading to lower clustering yield.

According to the above description, it should be expected that the nonlinear medium responsible for the observed HG, in both nanosecond and picosecond ranges, is constituted not only by metal atoms and ions but also by clustering species [19]. Cluster synthesis is often described as being carried out by laser ablation due to the rapid sample temperature increase, which favors plasma gas phase reactions [20, 21]. Condensation of gaseous metal atoms to small clusters can be achieved by adiabatic cooling of the plume expanding in vacuum and reactive collisions within the plasma volume. Higher laser pulse energies lead to a higher density plasma and thus to a higher number of collisions between the ablated atoms, resulting in more efficient condensation and production of larger size clusters.

With regard to the most versatile and efficient material for low-order HG of the fundamental 1064 nm radiation using the investigated ablation plasmas, the results presented here indicate that the nanosecond plume of manganese can sustain HG in a large spatiotemporal plasma region and in a wide range of ablation energies. Earlier studies of HHG in the laser ablation plasma of this medium have demonstrated an extremely high harmonic cutoff with generation of up to the 101st harmonic [14]. On the contrary, upon picosecond ablation, the active temporal and spatial regions of the plume are very much restricted with respect to nanosecond ablation confined to few tens of nanosecond after the ablation event and few hundred of micrometers close to the target. In both nanosecond and picosecond cases, these results give insight on the nature of the nonlinear emitters responsible for HG, although more definitive conclusions should be drawn to characterize the cluster and nanoparticle content of the metal ablation plumes by time-of-flight mass spectrometry and by analysis of the plume material deposited on a substrate placed in the proximity of the target [22, 23].

3.2 Low-Order Harmonic Generation in Nanosecond Laser Ablation Plasmas of Carbon-Containing Materials

In the case of low-order HG, two perspectives can be considered: on the one hand, it can be useful in the search for new efficient materials to act as the nonlinear media in HG processes. On the other hand, it can be considered as a diagnostic tool for the characterization of species present in the ablation plasma through the spatiotemporal study of the emitted harmonics [11–13, 24]. Optical emission spectroscopy (OES) is a well suited technique to investigate the composition of laser-induced plasmas; however, it usually provides information restricted to that derived from the emission of atoms and small molecules. Higher mass species, such as clusters and nanoparticles, could be also involved in the generation of harmonics, and therefore information regarding these plasma components can be obtained by the spatiotemporal analysis of harmonic signal. Moreover, this information is crucial to understand the HG process and increase the frequency up-conversion efficiency.

Low-order HG in nanosecond laser ablation plasmas of two carbon-containing materials, graphite and boron carbide (B_4C), has recently been analyzed in [25]. Laser ablation plasmas of carbon-containing materials have shown high harmonic conversion efficiency in the XUV range [26–32], and therefore, knowledge of the carbon plasma species responsible for the frequency up-conversion is of great interest. To this purpose, I discuss the results of systematic spatiotemporal study of the third ($\lambda = 355$ nm) and fifth ($\lambda = 213$ nm) harmonics of a 1064 nm driving laser in carbon-containing plasmas under different atmospheres [25]. OES and time-of-flight mass spectrometry (TOF-MS) analysis of the plumes of these two carbon-based materials, along with the analysis of deposits fabricated by pulsed laser deposition (PLD), gave further support to the conclusions obtained from the spatiotemporal study of HG. The results obtained have served to find a rich spatiotemporal distribution of nonlinear emitters of different sizes, ranging from small molecules to nanoparticles, in the plasma plumes of graphite and boron carbide.

The experimental set-up has been described in previous section, hence only a brief account is given here. The targets of graphite (purity 99.99%) and boron carbide (purity 99.9%) were mounted on a rotating holder to avoid crater formation on the surface and placed inside a vacuum chamber. Laser ablation plasmas were produced by the fundamental 1064 nm output of a Q-switched Nd:YAG laser (7 ns FWHM, 10 Hz) at normal incidence with respect to the target surface. The radiation was focused onto the target with a 17 cm focal length lens resulting in a spot size of 800 μm diameter. A fluence of 3 J/cm^2 was used for all experiments. A second Q-switched Nd:YAG laser (17 ns FWHM, 10 Hz) propagating parallel to the target surface and fired at a given temporal delay was used as a driving radiation. This radiation was focused in the plasma plume using a 30 cm focal length lens. The intensity in the focal area was nearly 0.5 GW/cm^2. In this arrangement, the directions of propagation of ablation laser and driving laser beams define the x and z axes, respectively. The system is disposed in such a way that the displacement of the target allowed the spatial scanning of the plume along the x axis (x-scan). In the same way, displacement of the driving laser focus position across the plasma plume allowed the spatial scanning along the z axis (z-scan). After the interaction of the driving pulse with the plasma, the TH radiation was guided to the entrance slit of a spectrograph and separated from the driving laser radiation with the help of two highly IR transmissive, UV reflective mirrors. The detection was performed with a time-gated ICCD camera with an integration time of 100 ns and by accumulation of the signal of 250 laser shots, corresponding to one target revolution. A second detection system was used for the detection of the generated FH. This system consisted of a photomultiplier coupled to a spectrograph. In OES experiments, the two dichroic mirrors used in HG measurements were replaced by an 8 cm focal length lens in order to form the image of the plume on the entrance slit of the detection system described for TH detection. Spectral acquisition was carried out at zero delay and by accumulation of the signal of the same number of laser shots. Experiments were carried out under vacuum conditions better than 2×10^{-2} mbar. In addition, a background pressure of 1 mbar of He, Kr, and Xe was used in the experiments performed under nonreactive gas atmosphere in order to assess the effect of aggregation of plasma species. TOF-MS analysis of the plume was carried out using the same experimental set-up as described in [32].

Material ejected in the ablation plume of the studied targets was collected on substrates, simultaneously with the HG measurements using a typical PLD arrangement [33]. The ablated material was collected under vacuum conditions on monocrystalline silicon (100) wafers, placed parallel to the target surface at a distance of 4 cm. Characterization of the deposits was performed by scanning electron microscopy.

A systematic study of the spatiotemporal behavior of the TH and FH generated in plasmas of graphite and boron carbide was performed by exploring the plasma plumes at controlled positions (x and z-scans) and for a range of temporal delays between the ablation event and the arrival of the driving laser pulse. As mentioned, the interest of studying carbon-containing materials resides in their high harmonic conversion efficiency in the XUV range. Graphite is a pure carbon compound that serves as a simple model in the study of HG processes. Boron carbide is an extremely hard chemical material used in tank armor, bullet-proof vests, and numerous industrial applications. With a hardness of 9.3 on the Moh's scale, it is one of the hardest materials known, behind cubic boron nitride and diamond. The use of such a sample is profitable from the point of view of stability of plasma formation (compared with milder graphite). It is especially important in the case of ablation with high pulse repetition rate lasers. On the contrary, in B_4C, boron atoms can contribute to the nonlinear process and differences in carbon content could affect the harmonics yield. The TH of the 1064 nm driving laser radiation at 355 nm was observed in both graphite and boron carbide. The nonlinear conversion efficiency was estimated for both materials following the procedure described in [11]. Boron carbide showed a three times higher TH conversion efficiency in comparison with the graphite plasma. The FH at 213 nm was also observed in case of B_4C plasma. The FH conversion efficiency was estimated to be one order of magnitude lower with respect to that of the TH in the case of boron carbide. In graphite, the detection of the weak FH signal was hindered by the overlapping intense atomic carbon emission lines in the region of 213 nm.

In order to confirm the energy dependence of the HG process in both materials, the evolution of the harmonics signal with the driving laser energy was studied. Figure 3.7 shows examples of those measurements for graphite and boron carbide. Values of the slopes in log-log plots are in agreement with the expected cubic dependence for the TH. The FH signal from the boron carbide plasma (not shown) displays a fifth power dependence as expected.

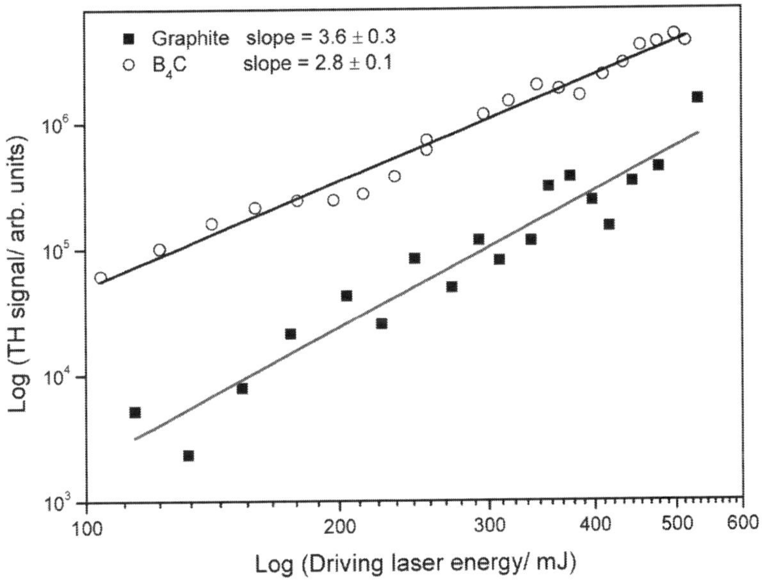

Figure 3.7 Log-log plot of the TH signal as a function of the driving laser energy in graphite (solid squares) and boron carbide open circles. Reproduced from Ref. 25, Copyright 2013, with permission from Elsevier.

The temporal evolution of the observed TH signal was analyzed by varying the delay between the ablation and the driving laser pulses. Figure 3.8 shows measurements in plasmas of graphite and boron carbide with the driving laser interacting with the plume at 0.6 mm above the target surface. The three panels of Fig. 3.8 correspond to the TH evolution under the different studied atmospheres. In case of inert gas atmospheres, an enhancement in the aggregation of plasma species as a result of multiple collisions with buffer gas atoms is expected. Previous studies have shown that the presence of a buffer gas favors the aggregation of species ejected in the ablation process [34–36]. On their flight, carbon species formed in the ablation undergo collisions yielding carbon clusters, which are stabilized by the transfer of the excess energy to the buffer gas atoms. Therefore, if heavier species, such as clusters and nanoparticles, are involved in the HG process, this contribution can be revealed. Under vacuum, in panel (a), a single maximum is observed in the TH emission at a delay of 350 and 250 ns for

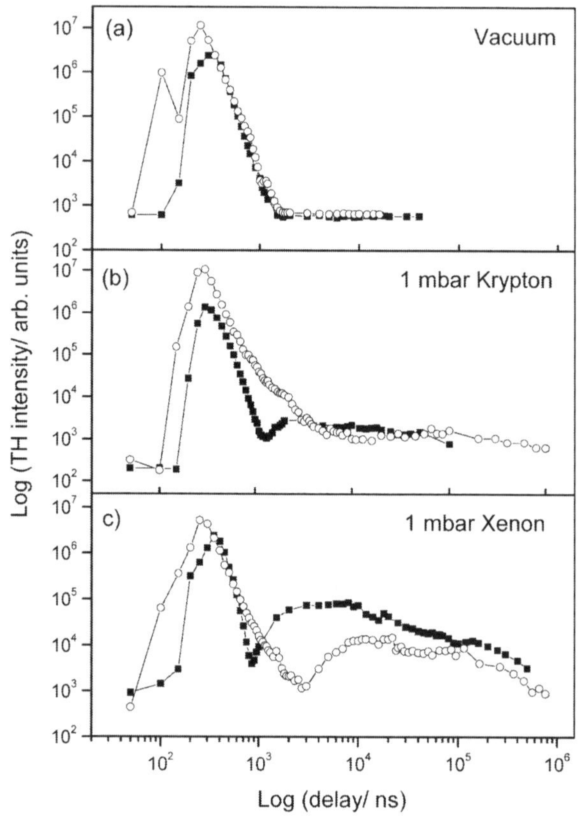

Figure 3.8 Log-log plot of the TH signal as a function of the delay between the ablation event and the arrival of the driving laser pulse at a distance of 0.6 mm from the target surface under different ambient conditions obtained in graphite (solid squares) and boron carbide (open circles): (a) vacuum, (b) 1 mbar of krypton, and (c) 1 mbar of xenon. Reproduced from Ref. 25, Copyright 2013, with permission from Elsevier.

graphite and boron carbide, respectively. This first component is attributed to lighter species that contribute to TH generation. Under 1 mbar of krypton and xenon, in panels (b) and (c), respectively, in Fig. 3.8, a second component is observed at longer delays for both graphite and boron carbide. This slower population of nonlinear species observed under buffer gas atmospheres can be attributed to heavier clusters and nanoparticles, which are responsible for the TH generation at longer delays. The reason for choosing inert gases with increasing atomic size is to evidence the influence of the collision diameter in the aggregation of species. The larger size of Xe atoms favors the probability of collisions and therefore the aggregation, resulting in a higher TH signal enhancement in comparison with that observed under Kr. On the contrary, this signal enhancement at longer delays due to the presence of heavier particles was not observed under helium atmosphere (not shown). As expected, increasing the buffer gas pressure leads to a higher degree of aggregation and therefore to an enhancement in the harmonic yield when performing the experiment at a long delay between the ablation event and the arrival of the driving laser pulse as a consequence of the contribution from heavier species.

Optical emission spectra of plasma plumes of the studied targets under the same conditions used in the HG experiments are shown in Fig. 3.9. In the case of graphite, in panel (a), the observed emissions are assigned to vibrational bands of the C_2 ($d^3\Pi_g \rightarrow a^3\Pi_u$) Swan system. At a shorter wavelength, the broad emission of C_3 is also observed [37]. Only when increasing the ablation fluence, atomic carbon emissions (neutral and ionic) made appearance in the spectrum. In the case of boron carbide, in panel (b), the acquired spectra show neutral emission lines from boron together with a weak signal corresponding to bands of the C_2 Swan system. This fact suggests that, in the selected ablation conditions, the main emitting species present in the graphite plume are mainly small molecules such as C_2 and C_3, while in the boron carbide plasma, emitting species are both atomic and molecular. In graphite plasma, a velocity of 2400 m/s was estimated for species responsible for TH generation. This value is in agreement with the velocity of excited C_2 molecules, about 3000 m/s, reported by Claeyssens et al. [18] in OES studies of 248 nm ablation plumes of graphite. These findings add support to the conclusion that C_2 and C_3 molecules are mainly responsible for the TH generation at short delays in the case of ablation using long (nanosecond) pulses.

TOF-MS measurements of the graphite plumes performed under vacuum better than 2×10^{-6} mbar, and under similar irradiation conditions as those used for HG, confirmed the presence of carbon clusters up to C_{30} [32]. On the contrary, and for further assessment

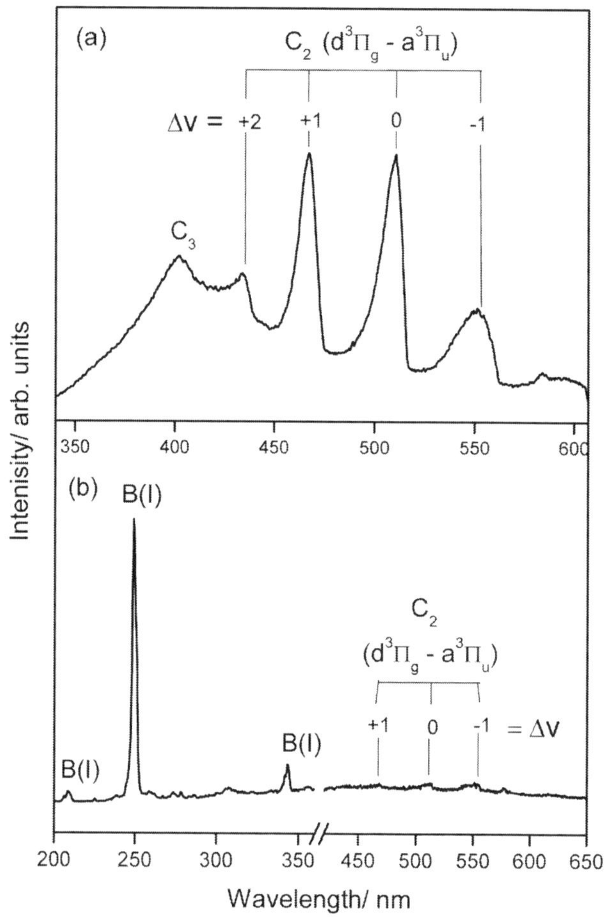

Figure 3.9 Optical emission spectra obtained by 1064 nm laser ablation at a fluence of 3 J/cm^2 under vacuum of: (a) graphite; (b) boron carbide. Assigned molecular and atomic emission bands are indicated. Spectra were obtained at zero delay after the ablation event and integrating the signal during 1 μs. Reproduced from Ref. 25, Copyright 2013, with permission from Elsevier.

of plasma composition, deposits fabricated by PLD were collected simultaneously with HG experiments. Deposits collected under vacuum appear highly uniform and the absence of particulates should be noted. A completely different situation was observed in case of deposits fabricated under 1 mbar of buffer gas. The deposits fabricated under Kr and Xe atmosphere showed characteristic dendritic structures of carbon grown by aggregation of clusters [34–36]. These results, with the appearance of a second component of TH at longer delays, confirm the presence of heavier species in the graphite plasma. Deposits collected by ablation of boron carbide targets equally reveal a nanostructured morphology.

3.3 Comparative Studies of Third Harmonic Generation in Plasma Plumes Using Picosecond and Femtosecond Laser Pulses

As it became obvious from previous consideration, the lowest order (i.e., third) HG in laser-produced plasma plumes [1, 3] is a rarely studied process compared with conventional THG in air and other gaseous media. THG in ionized air was analyzed beginning with the early stages of studies of laser–gas interaction by femtosecond pulses [38]. Various aspects of THG in air and other gases were analyzed in the following studies [39–46]. In the meantime, the advantages of in-plasma HG could largely be realized with the use of a low-excited and weakly ionized plasma produced on the surfaces of bulk targets. It is significant that the use of femtosecond pulses enabled producing lowest order harmonics from laser ablation with a relatively high conversion efficiency (of the order of 10^{-3}), which allows developing the recommendations for generation of shorter wavelength coherent radiation through the same nonlinear optical process. In this section, I analyze the studies of the THG in various plasmas using two lasers (Ti:sapphire and Nd:YAG) generating femtosecond and picosecond pulses [15]. These results are compared with the THG in air.

A Ti:sapphire laser (wavelength 795 nm, pulse duration 110 fs, pulse energy 8 mJ) at a 10 Hz pulse repetition rate was one of the laser sources used in these studies. In the case of THG studies in air, the laser radiation was focused inside a vacuum chamber by a 100 mm focal length lens. The beam waist radius of the focused radiation was 16 μm. Air pressure inside the chamber varied from 0.01 to 101 kPa. The spectral characteristics of third harmonic radiation (265 nm) were analyzed by a spectrometer and registered by a photomultiplier tube.

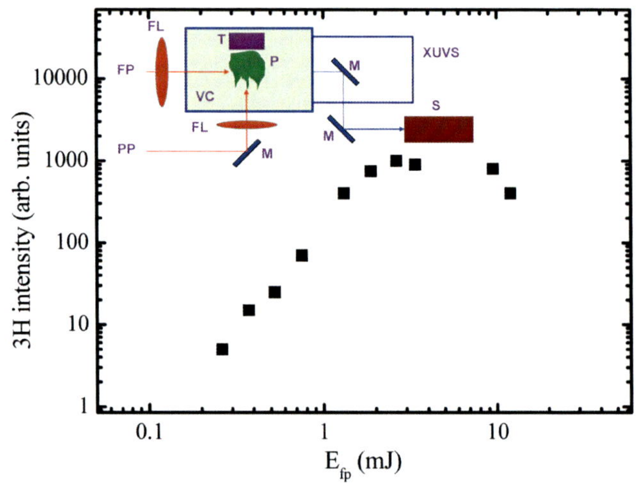

Figure 3.10 Third harmonic intensity as a function of fundamental pulse energy measured in the Mo plasma at a delay between pulses of 15 ns. Inset: Experimental setup for the HHG studies in laser plasma. FP: fundamental pulse; PP: pump pulse; FL: focusing lenses; T: target; VC: vacuum chamber; P: plasma; XUVS: XUV spectrometer; S: USB2000 spectrometer. Reproduced from Ref. 15 with permission from IOP Publishing.

The same laser was used in the THG studies in laser-produced plasmas. To create a plasma plume, a laser pulse was split from the amplified laser beam by a beam splitter before a compression of the fundamental pulse. The heating picosecond pulse duration was 210 ps. A spherical lens focused this beam on a target placed in a vacuum chamber to generate a laser ablation plume (see inset in Fig. 3.10). In these experiments, we used the B, Cr, C, Mg, W, and Mo bulk targets. The area of ablation was adjusted to be approximately 0.6 mm. The picosecond pulse intensity at the target surface was varied in the range of $I_{pp} = (0.5\text{–}5) \times 10^{10}$ W/cm^2. A compressed fundamental pulse at a center wavelength of $\lambda = 795$ nm had the energy of 12 mJ with a pulse duration of 110 fs after the propagation through the compressor stage. After the proper delay with regard to the pump pulse, this radiation was focused on the ablation plume from the orthogonal direction.

The lowest harmonic generated in metal plasmas was analyzed using the spectrometer USB2000 (Fig. 3.10). We installed the reflecting mirror in front of the grating in the XUV spectrometer. The harmonic and fundamental radiations were reflected through the window of the XUV spectrometer toward the spectrometer. Only the lowest (third order, $\lambda = 265$ nm) harmonic, together with fundamental radiation, was able to propagate through the fused silica window. We analyzed the spectrum and energy of this harmonic and optimized the third harmonic efficiency by varying the delay between pump and fundamental pulses, distance between the laser beam and target, intensities of the pump and fundamental pulses, focal position of the fundamental beam, and so on. The application of nonlinear crystals (β-barium borate, KDP) in front of the focusing lens allowed for the absolute calibration of THG efficiency. We measured the energy of the third harmonic generated in these crystals using the calibrated power meters. This third harmonic radiation then propagated through the vacuum chamber and laser plume, and registered by the spectrometer, thus giving the absolute calibration of our setup in the range of 265 nm [47].

The THG picosecond pulses were recorded using the experimental scheme described in Section 3.1. A brief account of this scheme is given below. A Nd:YAG laser with passive mode locking generated a train of pulses ($\lambda = 1064$ nm) of 50 ps duration. The separation between the pulses was 19 ns and the pulse train duration was 190 ns. We separated a single pulse from the central part of the pulse train using the Pockels cell placed inside the oscillator. A few first pulses before the separated pulse were used for plasma formation on the target surfaces placed in the vacuum chamber, which was maintained at different pressure of air. These picosecond pulses were focused by a 10 cm focal length lens on the Mn, B, or In targets. After some delay with regard to the last pulse of train, the fundamental single pulse was focused in the area of plasma plume. The focused radiation had the following characteristics: the energy of single pulse was 4 mJ, the pulse duration was 50 ps, and the intensity of a single pulse at the focus was varied up to 3×10^{13} W/cm^2. The third harmonic of fundamental laser radiation generated in the plasma was directed to a spectrometer. In the case of THG experiments in air, the same scheme was utilized, without the ignition of laser plasma on the surface of targets.

The THG in beryllium plasma was reported in [48] and the recommendations derived from those studies allowed for the enhancement of higher order harmonics. In [15], we analyzed the lowest HG in various plasmas using the femtosecond pulses, as well as in the Mn, B, and In plasmas using the picosecond pulses.

In the case of 110 fs pulses, the lowest harmonic generated in Mo and B plasmas was studied at different experimental conditions. The influence of third harmonic intensity on the 110 fs pulse energy showed a monotonic increase up to 3 mJ with further saturation and gradual decrease of HG efficiency (Fig. 3.10, molybdenum plasma). The slope of $I_{3\omega}(I_{1\omega})$ dependence was equal to 2.6. The maximum intensity of femtosecond laser beam at the focal spot could reach 8×10^{16} W/cm^2 at the maximum pulse energy available from the laser. As this intensity considerably exceeds the barrier suppression intensity for singly charged Mo and B ions, the position of the laser focus was adjusted to be either before or after the laser plume to optimize the third harmonic output. The intensity of the femtosecond laser pulse at the position of the preformed plasmas was varied between 2×10^{14} and 5×10^{15} W/cm^2.

The HG efficiency was optimized by choosing the optimal position of the focus of the fundamental laser beam relative to the plasma plume. It was observed that, for very intense laser pulses ($>10^{16}$ W/cm^2) focused in the center of the plume, the harmonic radiation was considerably decreased. However, the harmonic intensity showed a maximum when the fundamental laser beam was focused either before or after the plume, depending on the experimental conditions. We achieved the 1×10^{-3} and 4×10^{-4} conversion efficiencies of the third harmonic radiation in the boron and molybdenum plasmas, respectively, at the optimal position of laser focus. The analysis of third harmonic spectrum showed the unsaturated profile, which repeated the spectral distribution of fundamental radiation.

For the short delay (15 ns), the maximum harmonic output from the boron plasma at high intensities of the pump pulse on the surface of bulk target ($\sim 3 \times 10^{10}$ W/cm^2) was obtained by focusing after the plasma position (Fig. 3.11a). The opposite feature was observed for the 60 ns delay, when the maximum HG efficiency was achieved when the fundamental pulse was focused before the plasma. Such

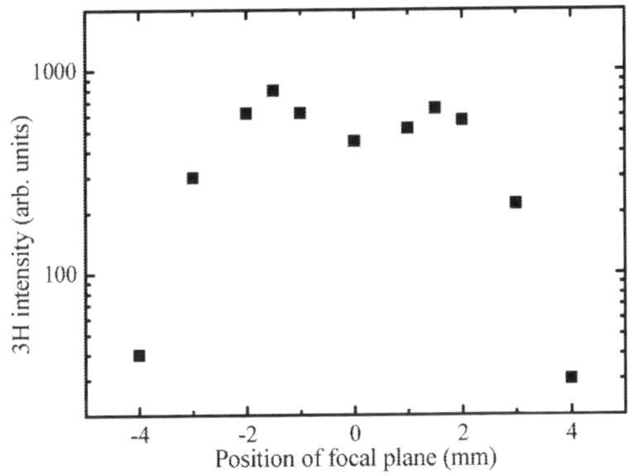

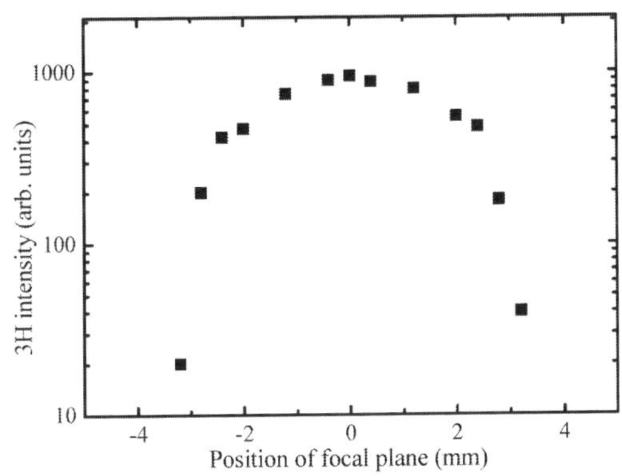

Figure 3.11 Dependences of third harmonic intensity on the position of the focal plane of focused 110 fs radiation relative to the position of plasma plume at (a) high ($\sim 3 \times 10^{10}$ W/cm^2) and (b) weak ($\sim 1 \times 10^{10}$ W/cm^2) intensity of the heating pulse on the target surface. Negative values on Fig. 3.11 correspond to the focusing after the plasma plume. Reproduced from Ref. 15 with permission from IOP Publishing. All rights reserved.

a behavior is similar to those reported by several authors in laser-gas jet HHG experiments [49–51], which was attributed to the depletion, phase-mismatching, and self-defocusing effects at high laser intensity.

We now examine the earlier results in terms of the physical picture of HG and the effect of laser focusing conditions. Low-order HG in gases or plasma plumes is well understood in terms of nonlinear interaction between the laser field and the bound electrons of the atoms/ions [16]. The high electric field of the ultrashort laser creates a strong distortion of the atomic field. The highly nonlinear motion of the bound electrons in the atom under the influence of the strong external electric field of the laser is responsible for low-order HG.

In the cases of laser plasma, the higher is the electric field (intensity) of the laser, the higher is the harmonic conversion. However, as the laser intensity increases, the already existing ions of plasma get increasingly over-ionized, thereby increasing the density of free electrons in the plasma and decreasing the population of singly charged ions. It has been shown in earlier HHG studies that the harmonic radiation generated in plasma plumes mainly comes from singly charged ions [52]. As the best focus intensity ($>10^{16}$ W/cm^2) is considerably larger than nearly 10^{15} W/cm^2 required for a complete ionization of singly charged ions to higher ionization species [53], the population of these ions is greatly depleted, which results in a reduced number of harmonic emitters. Moreover, the increased electron density enhances the phase mismatch between laser pulse and harmonic radiation. Hence, reduced singly charged ion density and increased electron density are responsible for the reduced harmonic emission. Note that, in the case of weak pump pulse ($\sim 1 \times 10^{10}$ W/cm^2), when the plasma concentration was maintained at the moderate level, the dependence of harmonic intensity on the position of the focal plane of focused radiation becomes monotonic, with maximum in the vicinity of plasma plume (Fig. 3.11b).

We analyzed the influence of the delay between pump and fundamental pulses on the efficiency of third harmonic. Figure 3.12a shows that initially, with increase of delay, the intensity of harmonic radiation increases. Further increase of delay leads to saturation and gradual decrease of third harmonic intensity. The optimal delay depends on the target material. To optimize the THG, we analyzed different targets. It was shown that lighter targets demonstrated higher conversion efficiency (Fig. 3.12b). These measurements were carried out at a 25 ns delay. One can expect that this delay becomes optimal for lighter neutrals and ions, as, for heavier particles, the time of flight from the target toward the interaction area becomes longer (due to smaller velocities compared with lighter particles).

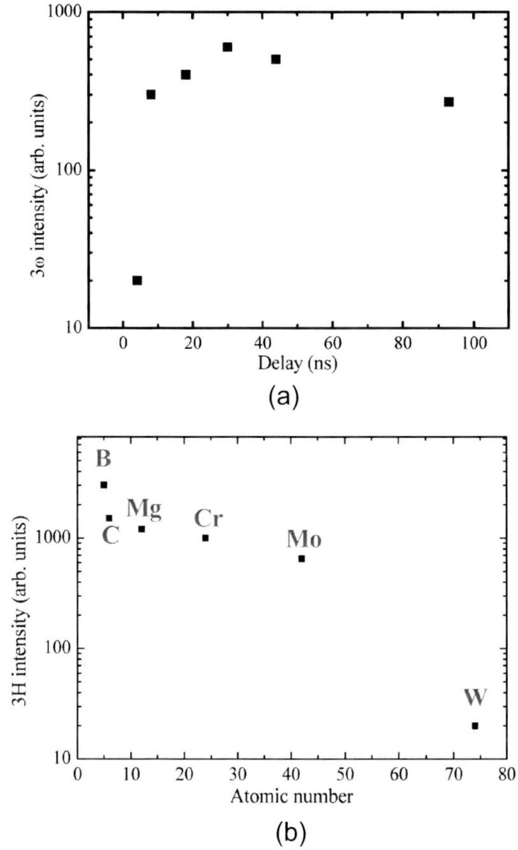

Figure 3.12 Dependence of third harmonic intensity on (a) the delay between 300 ps heating pulses and 110 fs fundamental pulses in the case of boron plasma and (b) atomic number. Reproduced from Ref. 15 with permission from IOP Publishing. All rights reserved.

In the case of picosecond pulses, we carried out the HHG in the laser plasma produced on the surfaces of different targets at two experimental conditions. Firstly, we analyzed the harmonic efficiency at ambient conditions, when the air pressure above the target surface was maintained at normal (atmospheric) conditions. Secondly, we generated the harmonic radiation when target was ablated at vacuum conditions.

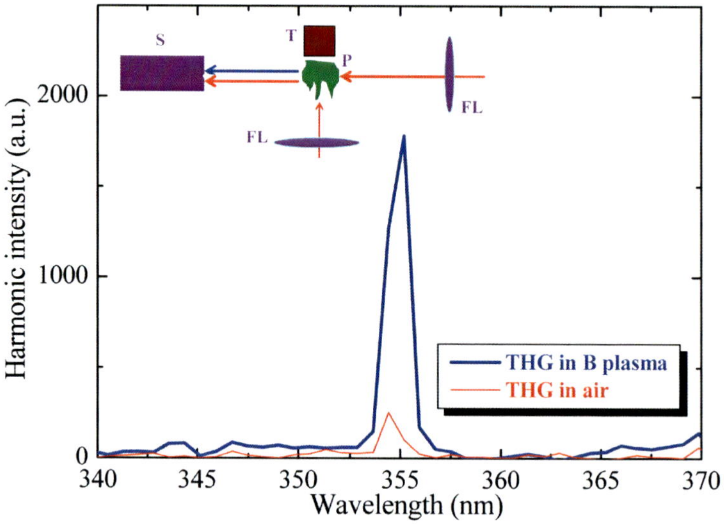

Figure 3.13 Third harmonic generation in boron plasma (thick curve) and air (thin curve) in the case of 50 ps driving pulses. Inset: Experimental setup for the THG in plasma using the picosecond pulses. FL: focusing lenses; P: plasma; T: target; S: spectrometer USB2000. Reproduced from Ref. 15 with permission from IOP Publishing. All rights reserved.

Our THG experiments demonstrated that plasma produced at the presence of air is a medium suitable for efficient conversion of coherent infrared radiation in the UV range and limited by the atmospheric transmission edge (200 nm). We shall now give the results of an investigation of the temporal, spectral, and energy characteristics of the generated radiation and of the plasma. Note that our studies of THG in Mn, B, and In plasmas at the conditions when plasma formation was accomplished at vacuum conditions did not demonstrate the enhancement of conversion efficiency compared with the case of plasma formation in air conditions.

Figure 3.13 presents the spectrum of the third harmonic generated in the boron plasma. For comparison, we present here the THG obtained in air (i.e., without the boron plasma). One can see a six-fold enhancement of harmonic efficiency in the former case. Analogous tendency was observed in the cases of indium and manganese plasma. The maximum THG efficiency ($\sim 10^{-4}$) has been

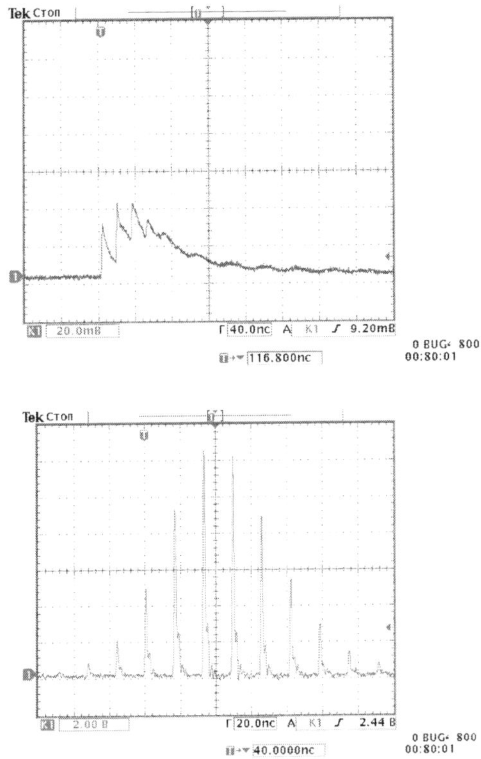

(a)
(b)

Figure 3.14 Temporal traces of (a) the indium plasma emission from the part of pulse train (one division is equal to 40 ns) and (b) entire pulse train (one division is equal to 20 ns). Reproduced from Ref. 15 with permission from IOP Publishing.

achieved in the conversion of infrared picosecond radiation to the third harmonic using the manganese plasma.

The analysis of oscilloscope traces of the train of pump pulses ($\lambda = 1054$ nm) and single output radiation pulse ($\lambda = 351$ nm) showed that efficient THG in indium plasma began after some delay. The variation of the delay between the last pulse of train and fundamental pulse has modified the conditions of the laser plasma interacting with ultrashort radiation. Each subsequent pump pulse created plasma starting from the third pulse of the pulse train. Every later heating pulse improved successively the conditions for the

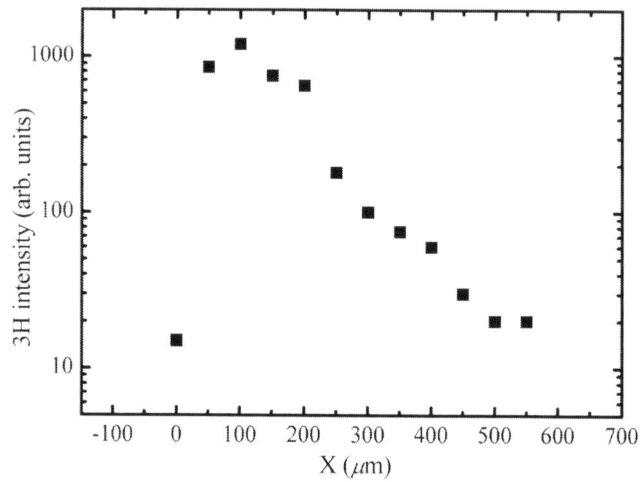

Figure 3.15 Dependence of third harmonic intensity on the distance between the 50 ps fundamental beam and target surface. Reproduced from Ref. 15 with permission from IOP Publishing. All rights reserved.

conversion of the fundamental radiation into harmonics because of the absorption in the plasma and its excitation. Figure 3.14 shows the oscilloscope traces of pulse train and plasma emission from the indium plasma. In the latter case, plasma formation began during interaction of the third or fourth pulses of the train with target surface. The best delay time for THG in these experiments was 25–40 ns.

We investigated the influence of the distance between the target and optical axis of propagation of the fundamental radiation. This distance was varied in the course of experiment by a manipulator, which controlled the position of the target relative to the waist of the fundamental radiation. Figure 3.15 demonstrates a considerable influence of this parameter on the nonlinear conversion to the third harmonic, which was associated with optimization of the plasma characteristics (plasma density and degree of excitation of higher states of the target plasma and atmospheric molecules) in the interaction zone. The optimal distance of propagation of the fundamental beam above the target surface was 100 μm.

Figure 3.16 shows the B and Mn plasma spectra in the UV/visible ranges at the conditions when the THG efficiency was maximal. One

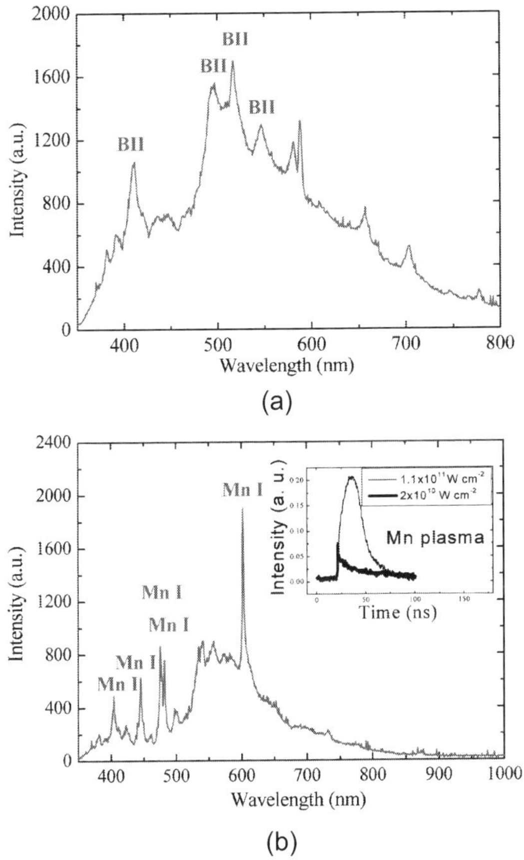

Figure 3.16 Plasma spectra from (a) boron and (b) manganese laser plumes. Inset shows the temporal trace of Mn plasma emission at two intensities of heating picosecond pulse on the surface of manganese target. Reproduced from Ref. 15 with permission from IOP Publishing.

can see that, in the case of light target (B), the plasma emission consisted of the transitions of ionic particles, although in the case of heavy target (Mn), we observed the emission from excited neutrals. We also analyzed the temporal shape of plasma emission from Mn plasma (see inset in Fig. 3.16b). At moderate excitation of target, the plasma emission considerably decreased after few nanoseconds, while in the case of strong excitation, plasma emission existed during few tens of nanoseconds. Optimization of target excitation is a crucial task when one needs to maximize the harmonic output from the plasma.

For comparison, we briefly describe our studies of THG in air (i.e., without the ignition of the plasma plume on the targets) using the femtosecond and picosecond radiation in tight focusing conditions. In the case of femtosecond pulses, the studies of THG as a function of fundamental intensity were carried out with radiation of different pulse duration. Femtosecond pulse duration was changed by variation of the distance between gratings in the compressor. In particular, for the 280 fs pulses, the experimental slope of the $I_3 \sim (I_1)^n$ dependence was close to $n \approx 2.7$ up to the fundamental intensity of $I_1 = 5 \times 10^{14}$ W/cm^2. Further growth of the fundamental intensity led to considerable decrease of the slope ($n \approx 1.3$). The optical breakdown caused by tunneling, multiphoton ionization, or both was observed at $I_1 = 4 \times 10^{14}$ W/cm^2, which was close to previously reported data (2.9×10^{14} W/cm^2 [54] and 2×10^{14} W/cm^2 [55]).

The point of slope change corresponded to the saturation intensity (I_s). Saturation of ionization is a typical effect that occurs during the multiphoton ionization experiments when the probability of ionization becomes close to 1, leading to the depletion of neutral atoms. The mechanism for interaction between the laser pulse and air changes considerably when the laser intensity exceeds I_s. For $I_1 < I_s$, the fundamental light interacts mostly with neutral molecules; when $I_1 > I_s$, the radiation interacts with ionized plasma. The increase of ion and electron concentration leads to considerable variation between the phases of fundamental and harmonic radiation owing to the strong influence of free electron induced dispersion. We found that small variations of laser intensity cause large fluctuations of TH emission in the intensity range 9×10^{13} to 3×10^{14} W/cm^2, whereas the fluctuations of THG efficiency at higher intensity become smaller. This fact underlines the strong dependence of this process on intensity at relatively low intensities and the decrease of this dependence at higher intensity.

THG in air using 1064 nm, 50 ps pulses has shown less conversion efficiency (2×10^{-5}) compared with the case of 790 nm, 110 fs pulses ($\sim 3 \times 10^{-4}$).

Below, we address the mechanism of plasma formation above the target surface. The creation of nonlinear medium above the target surface is not based on the simple heating of the target surface and steady-state processes of melting, evaporation, and spreading of the particles with the velocities defined from the thermodynamic relation. This relation refers to continuous waves heating. In this case, the velocity of the atoms at 1000 K is in the range of 7×10^2 m/s. During the first few nanoseconds ($\sim$10 ns), the atoms and ions will move only 7 μs above the surface. If one assumes that plasma creation by short laser pulse is defined by this slow process, then no harmonics at all should be observed in such experimental configuration for any target. For example, laser ablation of silver creates very efficient "nonlinear" plasma, which allows generating extremely strong high-order harmonics when the femtosecond pulse propagates 100 μm above the target surface [47]. The same can be said about any other targets, which was reported in multiple studies (see e.g., [14, 56, 57]). However, when one assumes a steady-state process of particle evaporation, the velocity of Ag neutrals and ions can be about 5.5×10^2 m/s. During first 30 ns, they travel only 16 μm away from the target surface, which is quite insufficient to reach the axis of the driving pulse.

This contradiction is explained by another model of creation of the cloud of particles, namely plasma explosion during ablation of the targets. The dynamics of plasma front propagation during laser ablation is studied by few groups (e.g., [58] and references therein). Numerical analysis of the generation of such plasmas for the case of single-pulse interaction with the target surface was performed in [59]. The dynamics of plasma formation and spreading can be analyzed by the shadowgraphs technique. Previously, the spatial characteristics of laser plasmas generated from B and Mo targets, measured using the shadowgraphs of the plasma, were reported in [60]. For heavy targets (Mo), the plasma front spreads with the velocity of 6×10^4 m/s. For example, the plasma front reaches 130 μm distance from the target after few nanoseconds (not few hundred nanoseconds, as is estimated if we assume a steady-state expansion of the plasma particle cloud). Obviously, the formation of "optimal" plasma is not restricted by appearance of the plasma front in the area of fundamental pulse propagation. One can wait until the density of the particles becomes sufficient for efficient HG, although the free electron concentration remains below the level when the impeding processes prevail over the HG.

The above assumptions and our experimental results show that optimization of the lowest harmonic allows us defining the conditions of plasma formation for efficient higher-order HG. Although the efficiency of THG using picosecond pulses still remains at the moderate level (due to small intensity of the laser beam and lower thresholds of multiphoton ionization of particles in the plasma), the efficient frequency conversion using femtosecond pulses ($\sim 10^{-3}$ in the boron plasma) demonstrates the applicability of proposed technique for generation of effective harmonics in the shorter wavelength range (100–200 nm) restricted by the transmission of optical materials. The indirect confirmation of this assumption was presented in [61], wherein the first ever observation of efficient HHG in laser plasma (up to the 61st order in the boron plume) was demonstrated. The same can be said about the manganese plasma. Earlier studies of this medium have demonstrated its suitability for highest ever harmonic cutoff observed in laser ablation (101st harmonic, [62]).

3.4 Low-Order Harmonic Generation of 1064 nm Radiation in Long Plasma Plumes

As it was already mentioned, a new history of plasma HHG studies has started in 2005 [61] when low-excited, low-ionized plasma plumes were applied for harmonics generation. It immediately followed with the observation of extended harmonics and considerably higher conversion efficiencies, which became comparable with those reported in gas HHG studies. During following few years, the enormous improvements in the characteristics of this process were reported. A substantial increase in the highest order of the generated harmonics, the emergence of a plateau in the energy distribution of harmonics, the high efficiencies obtained with several plasma formations, the realization of resonance-induced enhancement of individual harmonics, the efficient harmonic enhancement for the plasma plumes containing clusters of different materials, and other features have demonstrated the advantages of using specially prepared plasmas for the HHG [63, 64]. More details will be discussed in the following chapters.

All these studies were performed using narrow plasma plumes ($\leq$0.5 mm). The sizes of plasma plumes were defined by the focusing conditions of the spherical lenses used for ablation of the targets. In the meantime, one could expect that application of longer plasmas would further enhance the conversion efficiency due to quadratic dependence of the nonlinear optical response of medium on the length of laser-matter interaction. To create such long plasma plumes (say, one order of magnitude longer than previously used plasma plumes by using a cylindrical focusing of ablating beam), one has to carefully choose the conditions of plasma formation. The intensity of the heating pulse focused by cylindrical lens on the targets surface should be same as in the case of spherical focusing, which requires higher energies of this radiation, a nontrivial task even at present developments of femtosecond laser technique. The absorption of extended plasma can prevent the enhancement of harmonic yield from such a medium. Further, the phase mismatch can cause a deterioration of the optimal relations between the phases of interacting waves. Once the coherence length of this process becomes less than the sizes of plasma, the reverse down-conversion of harmonics can considerably decrease the harmonic yield. The presence of free electrons and excited ions in extended plasma plumes can cause the growth of the role of some impeding processes, such as Kerr effect, during propagation of converting pulse through the extended medium.

To analyze all these processes, one has to start with the lowest HG in the long plasmas. The easiness of 3H generation in laser-produced plasmas allows defining various impeding processes with sufficient accuracy, thus revealing the best conditions for lower order HG in extended plasmas, which could be further amended in the case of higher order HG.

Notice that all early studies of low-order HG in plasma plumes (see previous sections), as well as some recent studies [13, 24], were performed using ablation beams tightly focused on the targets and produced narrow plasma plumes. In this section, I analyze the properties of long (5 mm) plasma plumes for low-order HG

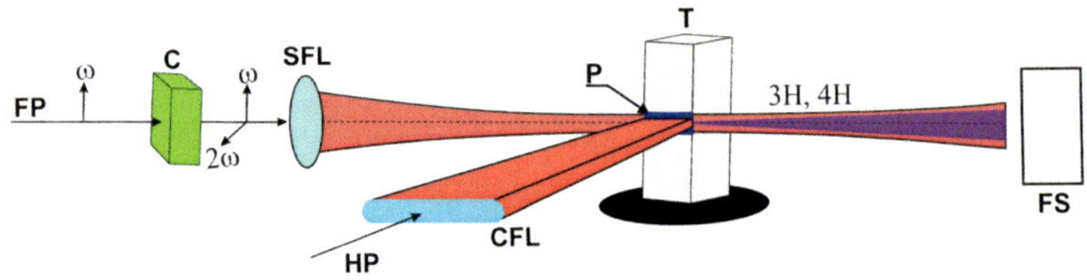

Figure 3.17 Experimental scheme. FP, fundamental pulse; HP, heating pulse; C, nonlinear crystal (KDP); SFL, spherical focusing lens; CFL, cylindrical focusing lens; T, target; P, plasma; FS, fiber spectrometer (HR4000). Reproduced from Ref. 65.

compared with the short (0.3–0.5 mm) lengths of plasmas used in previous studies. Third and fourth harmonics generation was studied using single-color and two-color pumps of the long plasmas produced on various metal surfaces. I also discuss the role of heating pulse characteristics for improvement of harmonic yield at these conditions of plasma formation [65].

The two-stage amplification of a single pulse of Nd:YAG laser (wavelength 1064 nm, pulse duration 38 ps) was followed by splitting of this radiation into two parts, one (heating pulse) with the energy of up to $E_{\text{hp}} = 15$ mJ, which was used for plasma formation on the target, and another (driving pulse) with the energy of up to $E_{\text{dp}} = 28$ mJ, which was used, after some delay, for frequency conversion in the prepared plasma. The heating pulse was focused using the 30 cm focal length cylindrical lens inside the vacuum chamber containing ablating target to create the line plasma plume (Fig. 3.17). The intensity of heating pulse on the target surface was 3×10^{10} W/cm^2. The driving pulse was focused on the prepared plasma from the orthogonal direction, parallel to the target surface. The confocal parameter ($b = 7$ mm) exceeded the length of plasma plume. The intensity of driving pulse at the focus was 4×10^{13} W/cm^2. The delay between these two pulses during most of experiments was maintained at 30 ns, which was optimal for efficient HG in laser-produced plasmas. The converted 3H radiations were analyzed using the spectrometer (HR4000).

To analyze the fourth harmonic (4H) generation, we used the two-color pump scheme. Part of driving pulse was converted to the second harmonic ($\lambda = 532$ nm) in the nonlinear crystal (KDP, type I, crystal length 10 mm) placed in front of the focusing lens. Conversion efficiency of second harmonic was 6%. Driving and second harmonic beams were focused inside the plasma plume. The polarizations of these two pumps were orthogonal to each other (Fig. 3.17).

The targets were made of various metals such as aluminum, tin, and copper. The width of targets where the ablation occurred was 5 mm. A three-coordinate manipulator made it possible to move the target and control the zone of the interaction of the driving radiation with the plasma relative to the target surface.

The important issue of these studies is a role of the length of plasma plume (d) in variation of the nonlinear optical response of this medium. One can expect various scenarios in the case of presence of the free electrons, which can drastically change the phase matching conditions between the driving and harmonic waves. The exceeding of the coherence length of 3H generation should lead to a decrease of harmonic yield. Once we assume that only micro-processes (related with the single particle response) rather than macro-processes (related with the collective propagation effects, and so on) govern the yield of harmonic emission, then the 3H intensity should follow the quadratic dependence on the length of nonlinear medium ($I_{3H} \propto d^l$, where $l = 2$ [16]).

We analyzed these dependences in a few plasma plumes. Figure 3.18 presents these results in the case of aluminum, tin, and copper plasmas. We changed the length of plasma using the slit placed after the cylindrical lens with the step of 0.5 mm. In the case of Al plasma (Fig. 3.18a), this dependence had a slope (l) close to 2.2 up to the $d \sim 2.5$ mm, with further insignificant decease of the slope ($l = 1.8$). The reasons for this change in $I_{3H}(d)$ dependence could be related with both micro-processes and macro-processes. Nevertheless, one can see a significant growth of 3H yield with the growth of plasma length. It means that free electrons, absorption of plasma, phase mismatch, Kerr effects, and so on did not play a decisive role in variation of this dependence at the optimal excitation of target.

3H generation in Sn (Fig. 3.18b) and Cu (Fig. 3.18c) plasmas showed analogous pictures, with less slope in the former case ($l = 1.7$) compared with copper plasma ($l = 2$). Thus, the tendency to

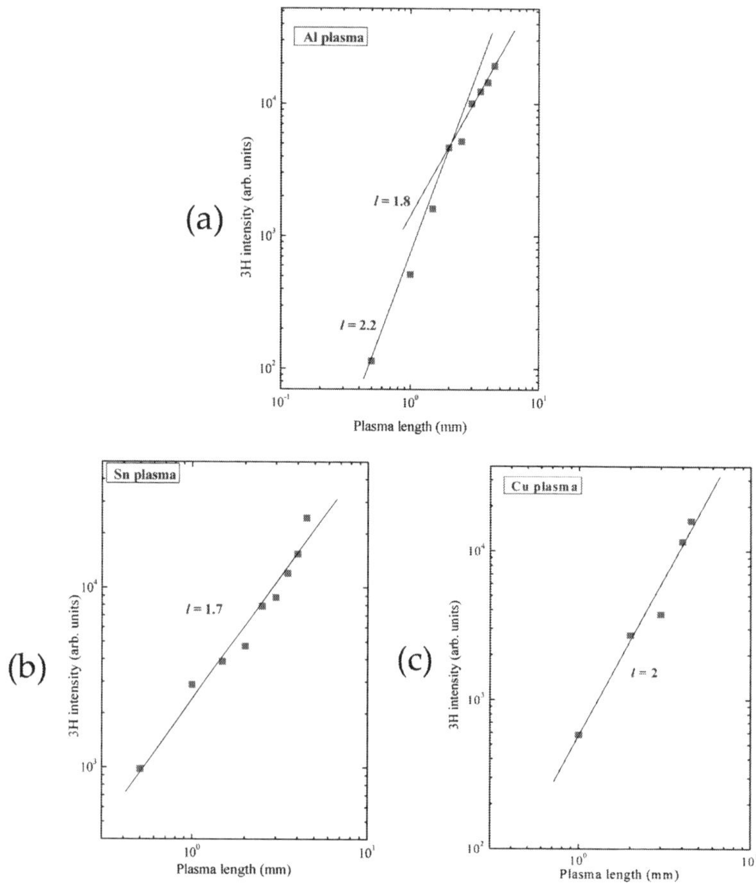

Figure 3.18 Dependence of the 3H intensity on the length of (a) Al, (b) Sn, and (c) Cu plasmas. Reproduced from [65].

follow the commonly accepted rule of harmonic-length dependence, at least for lowest order harmonic, seems a common feature at the conditions of our experiment (plasma concentration $\sim 2 \times 10^{18}$ cm^{-3}, heating pulse duration 38 ps, fluence on the target surfaces $\sim$1 J/cm^2).

The important parameter for optimization of low-order harmonic emission is the fluence of the heating pulse on the target surface. We measured the dependences of 3H efficiency on the energy of the heating pulse on the surfaces of Al and Sn, while

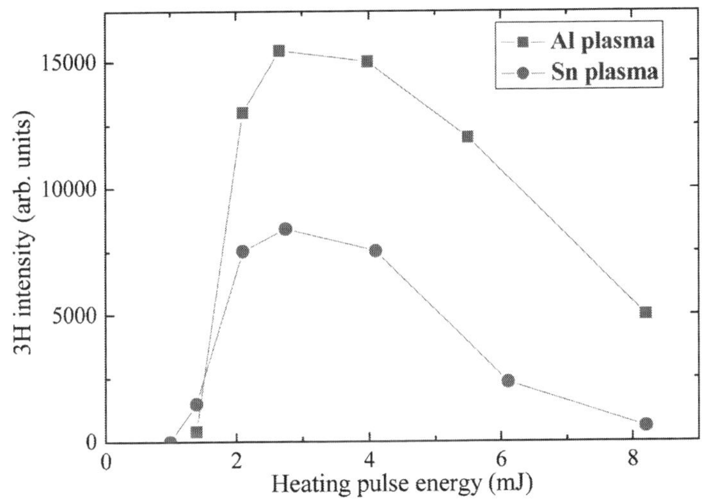

Figure 3.19 Dependence of the 3H intensity on the heating pulse energy in the cases of Al (filled squares) and Sn (filled circles) plasmas. Reproduced from [65].

maintaining the same geometry of ablation beam on the targets when the line focus from cylindrical lens created an extended plasma (5 mm). Figure 3.19 shows these dependences for 3H yields from the aluminum and copper plasmas. The common feature of these studies, which resembles those observed in the case of plasma HHG using picosecond heating pulses focused by spherical lens [66], was an observation of clearly defined maximum in $I_{3H}(E_{hp})$ dependence ($\sim$2.5 mJ for both cases) with following gradual decrease of harmonic at the irradiation of targets using stronger heating pulses. The reason of these observations is related with the over-excitation of the target, which leads to the appearance of the abundance of free electrons in the plasma plume, though in the case of 3H, this decrease of conversion efficiency with the growth of heating pulse intensity was not as abrupt as for higher orders harmonics.

4H generation using orthogonal driving fields has previously been reported only, to our best knowledge, in [67] wherein a simplified scheme was proposed for efficient 200 nm pulse generation in isotropic medium using 800 nm radiation, and the observation and optimization of the fourth-order HG during two-color filamentation in air were reported. This parametric process was based on four-wave mixing of the fundamental and the second-order harmonic radiation of Ti:sapphire laser in air plasma. A systematic study of the influence of the laser intensity, polarization, chirp, and pulse duration on the 4H output from the filaments has been carried out and, as a result of optimization of this process, the 4H conversion efficiency was estimated to be of the order of 10^{-4}.

It is well known that intensity inside the air filament can be maintained as high as 5×10^{13} W/cm^2. At these conditions, one can expect the creation of optimal phase-matching conditions for various parametric processes in air filaments, in particular 4H generation in the case of two-color laser (i.e., beam containing fundamental as well as second-order harmonic radiation). The same assumptions could be applied for the 4H generation in extended plasma plumes. The use of two-color pumping under these conditions may lead to an increase in the low-order HG efficiency similarly to previously published results on high-order harmonics in gases [68–73] and plasmas [74, 75] obtained using the 10 Hz and 1 kHz lasers.

The insertion of KDP crystal in front of spherical focusing lens led to second HG with efficiency of 6%. As it was shown in our studies, this ratio of second harmonic (532 nm) and main (1064 nm) driving pulses (1:16) was sufficient for observation of even (fourth) HG in extended plasma. Note that we did not observe this parametric process in the case of two-color pump when the plasma on the target surfaces was produced using spherical focusing of heating pulse and the sizes of plasma were of the order of few hundred micrometers. This observation point out the importance of long isotropic media for efficient 4H generation.

Figure 3.20 shows the typical spectral distribution of low-order harmonics in the case of the two-color pump of long aluminum plasma. In most cases, 4H was stronger than 3H, though the ratio between the intensities of these harmonics was unstable, due to some instabilities of laser and plasma formation. The conversion efficiencies of 3H and 4H were estimated as 3×10^{-5} and 5×10^{-5} using the approach similar to that described in [65] for calibration of high-order harmonics. Measurements of the absolute values of the conversion efficiencies of the 4H generated in the plasmas were

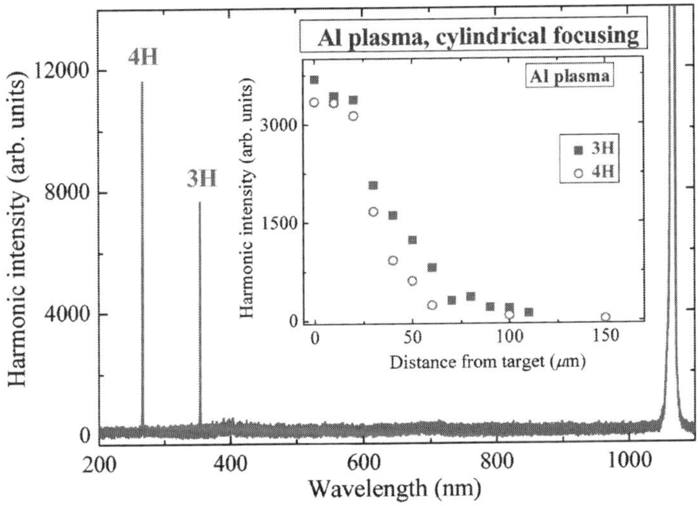

Figure 3.20 Spectra of radiation generating in extended Al plasma during two-color pump. Inset: Dependences of the 3H and 4H intensities on the distance between target surface and fundamental beam. Reproduced from [65].

carried out using the following procedure. In the first step, the 4H was measured by a spectrometer HR4000 using the known energy of the 4H of 1064 nm radiation generated in the nonlinear crystals. This allowed calibration of the spectrometer at a wavelength of 266 nm. The spectrometer allowed observation and measurement of the 3H and 4H conversion efficiencies.

The 3H and 4H intensities considerably depended on the distance between the optical axis of the driving beams and the target surface (see inset in Fig. 3.20). This dependence is caused by the change of plasma concentration above the target at a fixed delay between the heating and driving pulses. One can note that, in the case of spherical focusing of heating pulse, the analogous dependence showed HG up to the longer distances from the target (~100 and 200 μm for cylindrical and spherical focusing, respectively) at equal experimental conditions. This difference could be explained by different dynamics of plasma formation in these two cases. Note that the polarization of both 3H and 4H was followed by the polarization of driving 1064 nm and 532 nm beams, respectively.

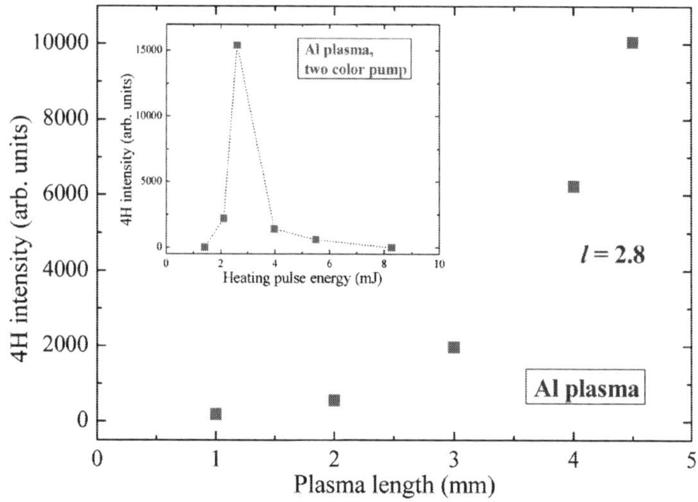

Figure 3.21 Dependence of the 4H intensity on the length of Al plasma. Inset: Dependence of the 4H intensity on the heating pulse energy in the cases of Al plasma. Reproduced from [65].

The nonlinear dependence of the 4H yield on the length of plasma medium was analyzed in the case of Al plasma plumes (Fig. 3.21). The slope of $I_{4H} \propto d^l$ dependence was steeper than the 3H case ($l = 2.8$). The reasons of the increase of a slope of this dependence are not clear, as both these processes are related with four-photon processes. The possible explanation of this difference is the involvement of macro-processes in overall yield of 4H.

The dependence of the 4H efficiency on the energy of heating pulse on the surface of aluminum target is presented in the inset to Fig. 3.21. We observed a sharp increase of 4H efficiency at $E_{hp} = 2.7$ mJ with following steep decrease of harmonic yield, contrary to the case of 3H (Fig. 3.19). The same strong dependence was observed in Sn plasma. This difference in $I_{harm}(E_{hp})$ dependences for 3H and 4H could have the same origin as above-described $I_{harm} \propto d^l$ dependences. Particularly, the propagation processes, which are related with the phase difference between the waves of two pumps and 4H, can play an important role in this parametric process. Some lack of spatial (and partially temporal) overlapping of two pumps (1064 and 532 nm) can vary the above two dependences and create the conditions when the optimization of 4H process occurs at a narrow range of the energies of heating pulses.

The insignificant difference in the low-order nonlinear optical properties of Al, Sn, and Cu plasmas demonstrates that these samples have no specific features in the near ultraviolet range, neither absorptive, nor resonance-induced ones. It means that plasma HG can be served as a precise tool for the nonlinear spectroscopy analysis of the atomic physics and structure of materials, which has been also proven earlier during analysis of the resonance-induced enhancement of single harmonics during HHG using different driving laser pulses.

In the case of long plasma, at the concentrations of plasmas of the range of 2×10^{18} cm^{-3} at the used intensities of heating pulses, the absorption can play rather important role compared with narrow plasma plumes. This difference should be emphasized for higher orders of harmonics due to larger cross-sections of absorption of ionic media in the shorter wavelength range.

Regarding the finding of these studies–stronger fourth harmonic compared with the third one–one can note that observation of stronger even harmonic compared with odd harmonic has already been observed in the case of HHG in some plasmas, while maintaining approximately same ratio between the fundamental and second harmonic driving fields (see [75] wherein the cutoff harmonics from two-color pump of Ag plasma consisted mostly of the even orders).

The use of long plasmas can give the opportunity in observation of quasi-phase matching (QPM) between the waves of driving pulses and harmonic pulses. For this, one has to create a bunch of plasma plumes of the sizes of coherence lengths for some specific spectral ranges. These lengths are significantly dependent on the dispersion of plasma, which can be adjusted by heating pulse fluence and variation of free electrons concentration. Moreover, once the long plasma-induced low-order HG becomes optimized, the following studies in the shorter wavelength range could reveal some attractive properties of the high-order harmonics generating at these conditions, such as enhancement of harmonic yield and creation of QPM conditions for some groups of harmonics in the XUV range. The application of relatively long picoseconds pulses at these conditions allows obtaining further ways for enhancement of harmonic yield.

References

1. S. M. Gladkov, N. I. Koroteev, *Sov. Phys. Uspekhi* 33, 554 (1990).
2. S. M. Gladkov, A. B. Fedotov, N. I. Koroteev, A. M. Zheltikov, *Sov. Tech. Phys. Lett.* 14, 610 (1988).
3. A. B. Fedotov, N. I. Koroteev, A. M. Zheltikov, *Laser Phys.* 5, 835 (1995).
4. A. B. Fedotov, S. M. Gladkov, N. I. Koroteev, A. M. Zheltikov, *J. Opt. Soc. Am. B* 8, 363 (1991).
5. C. Y. Ng, *Ann. Rev. Phys. Chem.* 53, 101 (2002).
6. A. H. Zanganeh, J. H. Fillion, J. Ruiz, M. Castillejo, J. L. Lemaire, N. Shafizadeh, F. Rostas, *J. Chem. Phys.* 112, 5660 (2000).
7. X. F. Li, A. L'Huillier, M. Ferray, L. A. Lompre, G. Mainfray, *Phys. Rev. A* 39, 5751 (1989).
8. C. Spielmann, N. H. Burnett, S. Sartania, R. Koppitsch, M. Schnurer, C. Kan, M. Lenzner, P. Wobrauschek, F. Krausz, *Science* 278, 661 (1997).
9. N. Hay, R. de Nalda, T. Halfmann, K. J. Mendham, M. B. Mason, M. Castillejo, J. P. Marangos, *Eur. Phys. J. D* 14, 231 (2001).
10. N. Hay, R. Velotta, M. B. Mason, M. Castillejo, J. P. Marangos, *J. Phys. B: At. Mol. Opt. Phys.* 35, 1051 (2002).
11. M. Oujja, R. de Nalda, M. López-Arias, R. Torres, J. P. Marangos, M. Castillejo, *Phys. Rev. A* 81, 043841 (2010).
12. R. de Nalda, M. López-Arias, M. Sanz, M. Oujja, M. Castillejo, *Phys. Chem. Chem. Phys.* 13, 10755 (2011).
13. M. López-Arias, M. Oujja, M. Sanz, R. A. Ganeev, G. S. Boltaev, N. K. Satlikov, R. I. Tugushev, T. Usmanov, M. Castillejo, *J. Appl. Phys.* 111, 043111 (2012).
14. R. A. Ganeev, *J. Phys. B: At. Mol. Opt. Phys.* 40, R213 (2007).
15. R. A. Ganeev, G. S. Boltaev, R. I. Tugushev, T. Usmanov, M. Baba, H. Kuroda, *J. Opt.* 12, 055202 (2010).
16. J. F. Reintjes, *Nonlinear Optical Parametric Processes in Liquids and Gases*, Academic Press, Orlando (1984).
17. W. Theobald, C. Wülker, F. P. Schäfer, B. N. Chichkov, *Opt. Commun.* 120, 177 (1995).

18. F. Claeyssens, M. N. R. Ashfold, E. Sofoulakis, C. G. Ristoscu, D. Anglos, C. Fotakis, *J. Appl. Phys.* 91, 6162 (2002).
19. M. B. Knickelbein, *Phys. Rev. Lett.* 70, 014424 (2004).
20. H. Lillich, J. Wolfrum, V. Zumbach, L. E. Aleandri, D. J. Jones, J. Roziere, P. Albers, K. Seibold, A. Freund, *J. Phys. Chem.* 99, 12413 (1995).
21. U. Chakravarty, P. A. Naik, C. Mukherjee, S. R. Kumbhare, P. D. Gupta, *J. Appl. Phys.* 108, 053107 (2010).
22. M. Sanz, R. de Nalda, J. F. Marco, J. G. Izquierdo, L. Bañares, M. Castillejo, *J. Phys. Chem. C* 114, 4864 (2010).
23. M. Sanz, M. López-Arias, J. F. Marco, R. de Nalda, S. Amoruso, G. Ausanio, S. Lettieri, R. Bruzzese, X. Wang, M. Castillejo, *J. Phys. Chem. C*, 115, 3203 (2011).
24. M. López-Arias, M. Oujja, M. Sanz, R. de Nalda, R. A. Ganeev, M. Castillejo, *Mol. Phys.* 110, 1651 (2012).
25. I. Lopez-Quintas, M. Oujja, M. Sanz, M. Martín, R. A. Ganeev, M. Castillejo, *Appl. Surf. Sci.* 278, 33 (2013).
26. R. A. Ganeev, M. Suzuki, M. Baba, H. Kuroda, *J. Opt. Soc. Am. B* 22, 1927 (2005).
27. R. A. Ganeev, L. B. Elouga Bom, J. Abdul-Hadi, M. C. H. Wong, J. P. Brichta, V. R. Bhardwaj, T. Ozaki, *Phys. Rev. Lett.* 102, 013903 (2009).
28. L. B. Elouga Bom, Y. Pertot, V. R. Bhardwaj, T. Ozaki, *Opt. Express* 19, 3077 (2011).
29. Y. Pertot, L. B. Elouga Bom, V. R. Bhardwaj, T. Ozaki, *Appl. Phys. Lett.* 98, 101104 (2011).
30. R. A. Ganeev, P. A. Naik, H. Singhal, J. A. Chakera, M. Kumar, M. P. Joshi, A. K. Srivastava, P. D. Gupta, *Phys. Rev. A* 83, 013820 (2011).
31. R. A. Ganeev, T. Witting, C. Hutchison, F. Frank, P. V. Redkin, W. A. Okell, D. Y. Lei, T. Roschuk, S. A. Maier, J. P. Marangos, J. W. G. Tisch, *Phys. Rev. A* 85, 015807 (2012).
32. R. A. Ganeev, C. Hutchison, T. Witting, F. Frank, W. A. Okell, A. Zaïr, S. Weber, P. V. Redkin, D. Y. Lei, T. Roschuk, S. A. Maier, I. López-Quintás, M. Martín, M. Castillejo, J. W. G. Tisch, J. P. Marangos, *J. Phys. B: At. Mol. At. Phys.* 45, 165402 (2012).
33. M. Walczak, M. Oujja, J. F. Marco, M. Sanz, M. Castillejo, *Appl. Phys. A* 93, 735 (2008).
34. P. M. Ossi, C. E. Bottani, A. Miotello, *Thin Solid Films* 482, 2 (2005).
35. P. M. Ossi, A. Miotello, *J. Non-Cryst. Solids* 353, 1860 (2007).
36. W. An, X. Zaho, Z. Zhang, R. Su, *Appl. Surf. Sci.* 256, 2304 (2010).

37. L. Nemes, A. M. Keszler, C. G. Parigger, J. O. Hornkohl, H. A. Michelsen, V. Stakhursky, *Appl. Opt.* 46, 4032 (2007).
38. A. L'Huillier, L. A. Lompre, M. Ferray, X. F. Li, G. Mainfray, C. Manus, *Europhys. Lett.* 5, 601 (1988).
39. S. Baskus, J. Peatross, Z. Zeek, A. Rundquist, G. Taft, M. M. Murnane, H. C. Kapteyn, *Opt. Lett.* 21, 665 (1996).
40. A. B. Fedotov, N. I. Koroteev, M. M. T. Loy, X. Xiao, A. M. Zheltikov, *Opt. Commun.* 133, 587 (1997).
41. J. Peatross, S. Baskus, J. Zhou, M. M. Murnane, H. C. Kapteyn, *J. Opt. Soc. Am. B* 15, 186 (1998).
42. G. Marcus, A. Zigler, Z. Henis, *J. Opt. Soc. Am. B* 16, 792 (1999).
43. A. B. Fedotov, A. N. Naumov, V. P. Silin, S. A. Uryupin, A. M. Zheltikov, A. P. Tarasevitch, D. von der Linde, *Phys. Lett. A* 271, 407 (2000).
44. N. Aközbek, A. Iwasaki, A. Becker, M. Scalora, S. L. Chin, C. M. Bowden, *Phys. Rev. Lett.* 89, 143901 (2002).
45. H. Yang, J. Zhang, J. Zhang, L. Z. Zhao, Y. J. Li, H. Teng, Y. T. Li, Z. H. Wang, Z. L. Chen, Z. Y. Wei, J. X. Ma, W. Yu, Z. M. Sheng, *Phys. Rev. E* 67, 015401 (2003).
46. R. A. Ganeev, M. Suzuki, M. Baba, H. Kuroda, I. A. Kulagin, *Appl. Opt.* 45, 748 (2006).
47. R. A. Ganeev, M. Baba, M. Suzuki, H. Kuroda, *Phys. Lett. A* 339, 103 (2005).
48. R. A. Ganeev, M. Suzuki, M. Baba, H. Kuroda, *J. Opt. Soc. Am. B* 25, 2096 (2008).
49. A. L'Huillier, P. Balcou, *Phys. Rev. Lett.* 70, 774 (1993).
50. M. Bellini, C. Corsi, M. C. Gambino, *Laser Part. Beams* 20, 277 (2002).
51. A. Bouhal, P. Salières, P. Breger, P. Agostini, G. Hamoniaux, A. Mysyrowicz, A. Antonetti, R. Constantinescu, H. G. Muller, *Phys. Rev. A* 58, 389 (1998).
52. R. A. Ganeev, M. Suzuki, M. Baba, H. Kuroda, *Appl. Phys. B* 81, 1081 (2005).
53. Z. Chang, A. Rundquist, H. Wang, M. M. Murnane, H. C. Kapteyn, *Phys. Rev. Lett.* 79, 2967 (1997).
54. Y.-D. Qin, H. Yang, C.-J. Zhu, Q. Gong, *Appl. Phys. B* 71, 581 (2000).
55. X. Liu, D. Umstadter, E. Esarey, A. Ting, *IEEE Trans. Plasma Sci.* 21, 90 (1993).
56. S. Kubodera Y. Nagata, Y. Akiyama, K. Midorikawa, M. Obara, *Phys. Rev. A* 48, 4576 (1993).
57. C.-G. Wahlström, S. Borgström, J. Larsson, S.-G. Pettersson, *Phys. Rev. A* 51, 585 (1995).

58. B. Rus, P. Zeitoun, T. Mocek, S. Sebban, M. Kálal, A. Demir, G. Jamelot, A. Klisnick, B. Králiková, J. Skála, G. J. Tallents, *Phys. Rev. A* 56, 4229 (1997).
59. H. Hora, *Plasmas at High Temperature and Density*, Heidelberg, Springer (1991).
60. R. A. Ganeev, M. Suzuki, M. Baba, H. Kuroda, *Opt. Spectrosc.* 99, 1000 (2005).
61. R. A. Ganeev, M. Suzuki, M. Baba, H. Kuroda, T. Ozaki, *Opt. Lett.* 30, 768 (2005).
62. R. A. Ganeev, L. B. Elouga Bom, J.-C. Kieffer, T. Ozaki, *Phys. Rev. A* 76, 023831 (2007).
63. R. A. Ganeev, *Laser Phys.* 22, 1177 (2012).
64. G. S. Boltaev, R. A. Ganeev, I. A. Kulagin, N. K. Satlikov, T. Usmanov, *Quantum Electron.* 42, 899 (2012).
65. R. A. Ganeev, G. S. Boltaev, N. K. Satlikov, R. I. Tugushev, T. Usmanov, *J. Photon. Optoelectron.* 2, 96 (2013).
66. R. A. Ganeev, G. S. Boltaev, N. K. Satlikov, I. A. Kulagin, T. Usmanov, *J. Opt. Soc. Am. B* 29, 3286 (2012).
67. R. A. Ganeev, H. Singhal, P. A. Naik, J. A. Chakera, M. Kumar, P. D. Gupta, *Phys. Rev. A* 82, 043812 (2010).
68. E. Cormier, M. Lewenstein, *Eur. Phys. J. D* 12, 227 (2000).
69. I. J. Kim, C. M. Kim, H. T. Kim, C. H. Lee, Y. S. Lee, J. Y. Park, D. J. Cho, C. H. Nam, *Phys. Rev. Lett.* 94, 243901 (2005).
70. J. Mauritsson, P. Johnsson, E. Gustafsson, A. L'Huillier, K. J. Schafer, M. B. Gaarde, *Phys. Rev. Lett.* 97, 013001 (2006).
71. T. Pfeifer, L. Gallmann, M. J. Abel, D. M. Neumark, S. R. Leone, *Opt. Lett.* 31, 975 (2006).
72. D. Charalambidis, P. Tzallas, E. P. Benis, E. Skantzakis, G. Maravellas, L. A. A. Nikolopoulos, A. P. Conde, G. D. Tsakiris, *New J. Phys.* 10, 025018 (2008).
73. I. J. Kim, G. H. Lee, S. B. Park, Y. S. Lee, T. K. Kim, C. H. Nam, T. Mocek, K. Jakubczak, *Appl. Phys. Lett.* 92, 021125 (2008).
74. R. A. Ganeev, H. Singhal, P. A. Naik, I. A. Kulagin, P. V. Redkin, J. A. Chakera, M. Tayyab, R. A. Khan, P. D. Gupta, *Phys. Rev. A* 80, 033845 (2009).
75. R. A. Ganeev, C. Hutchison, A. Zaïr, T. Witting, F. Frank, W. A. Okell, J. W. G. Tisch, J. P. Marangos, *Opt. Express* 20, 90 (2012).

Chapter 4

High-Order Harmonic Generation in Plasma Plumes Using Picosecond Pulses

Earliest observations of the HHG in gases were reported using the picosecond lasers [1–3]. The harmonics from different gases up to the 21st and 33rd orders of 1064 nm radiation were reported at the intensities of 3×10^{13} W/cm^2, which led to the enormous growth of the interest in this area of nonlinear optics. The studies described in this chapter demonstrate that application of ablated plasma as a nonlinear medium can further amend this HHG technique using picosecond driving pulses, once the atoms, ions, and clusters become involved as the efficient sources of harmonic generation.

4.1 Harmonic Generation of Picosecond Nd:Yag Laser Radiation in Metal Ablation-Produced Plasmas

In isotropic media, the HHG using femtosecond laser pulses allows easily production of the coherent radiation in the XUV range. During the last 25 years, predominantly rare gases were employed as the target media for the HHG, which, however, imposed some physical and practical limits on the performance of the coherent XUV sources.

Plasma Harmonics
Rashid A. Ganeev

ISBN 978-981-4463-30-0 (Hardcover), 978-981-4463-31-7 (eBook)
www.panstanford.com

So far, only low conversion efficiencies of HHG have been reported using the gases as the nonlinear media, despite the enormous efforts. For practical applications of high-order harmonic sources, an increase in the photon flux would be beneficial. The generation of high-order harmonics in other isotropic medium, laser-produced plasmas, being for this purpose a relatively new medium, promises to yield these expectations, as well as to open the door for new developments in the laser–matter interaction.

Previous plasma HHG studies were carried out using the femtosecond pulses. The spectral range of those experiments was restricted at the longer wavelength side of nearly 80 nm, which was defined by the registration properties of commonly used detectors of harmonic spectra (microchannel plates). No studies of plasma HHG were performed in the spectral region above this wavelength. A search of resonance enhancement of single harmonic at these conditions can be justified by a presence of strong ionic and neutral transitions in the longer wavelength vacuum ultraviolet range (80–200 nm). Next, the application of longer pulses could avoid the impeding processes restricting the efficiency of harmonic generation in this region. As it was mentioned in previous chapters, there are two main limiting processes, which can decrease conversion efficiency in the ionic medium. The first one is an excess of free electrons concentration caused by over-excitation of target surface. The appearance of considerable amount of free electrons is a result of ionization of both neutral atoms and singly charged particles. This follows with self-defocusing of propagating probe pulse. The second process is directly related with the first one. The phase matching conditions of the driving and harmonic waves, which are maintained at the moderate amount of the free electrons appearing during ionization of neutrals, break up due to the growth of free electrons concentration. This increase occurs abruptly once the fluence of heating pulse reaches the threshold level, when the over-excitation of targets leads to appearance of doubly and higher charged particles. Further, the use of longer pulses can also increase the fluence of harmonic emission at comparable HHG conversion efficiencies from picosecond and femtosecond sources due to higher pulse energies available in the former case.

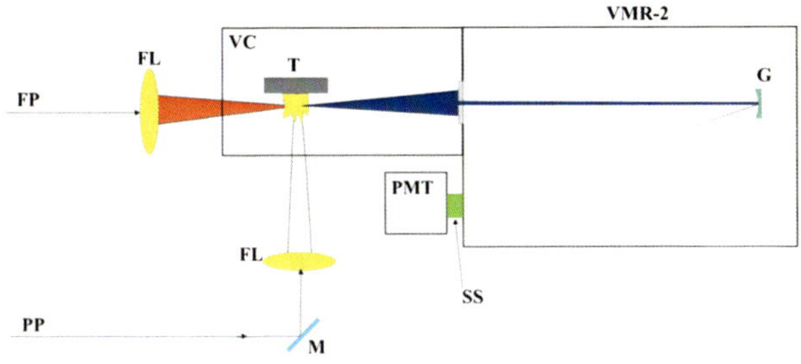

Figure 4.1 Experimental setup for the HHG in laser plasma using the picosecond pulses. FP, fundamental probe picosecond pulse; PP, heating picosecond pulse; M, mirror; FL, focusing lenses; VC, vacuum chamber; T, target; VMR-2, vacuum monochromator; G, grating; SS, sodium salicylate; PMT, photomultiplier tube. Reproduced from Ref. 4 with permission from IOP Publishing. All rights reserved.

The studies discussed in previous chapter have shown the attractiveness of application of the long laser pulses for third-order harmonic generation from the laser plasmas. In this section, I present the analysis of the HHG in various plasmas using the 1064 nm, 38 ps pulses [4]. The goal of these studies was a search of the conditions for generation of relatively high-energy coherent picosecond pulses in the range of 80–220 nm using various metal-ablated plasma plumes.

The passive mode-locked Nd:YAG laser ($\lambda = 1064$ nm, 1.5 Hz pulse repetition rate) generated a 38 ps pulse. The two-stage amplification of single pulse was followed by splitting of this radiation into two parts, one (heating pulse) with the energy of 5 mJ, which was used for plasma formation on the target surface, and another (probe pulse) with the energy of up to 28 mJ, which was used after some delay for frequency conversion in various plasma plumes (Fig. 4.1). The heating pulse was focused using the 300 mm focal length lens inside the vacuum chamber containing various targets. The plasma sizes were nearly 0.5 mm. The heating pulse intensity on the target surface was in the range of 10^{11} W/cm^2. The diameter of single-mode driving beam was 4 mm. This beam was focused inside the plasma plume using the 150 mm focal length lens. The driving beam propagated at the distance of 100–150 μm from the surface of targets. The intensity of probe pulse at the focus was

4×10^{13} W/cm^2. The focal spot was inside the plasma plume. The delay between heating and probe pulses during most of experiments was maintained at 25 ns, which was optimal for efficient harmonic generation in metal-containing plasmas. The harmonic radiation was analyzed using a vacuum monochromator.

Various metal materials (copper, chromium, zinc, niobium, silver, indium, molybdenum, titanium, tin, lead, tantalum, manganese, tungsten, gold, boron, and aluminum) were used as the targets for laser-induced plasma formation. A three-coordinate translating stage allowed the movement of the target along the z-axis and the control of the interaction zone of the driving radiation with the plasma relative to the target plane.

The analysis of the evolution of the spectra for diagnosing laser plasma provides important information about the plasma parameters and can be used for multiple applications. In particular, some time-resolved laser-induced plasma spectrometry studies during harmonic studies from the gold, silver, manganese, and vanadium plasmas [5] allowed the identification of the emissions from native species and optimization of conditions when plasma mostly consists of the excited neutrals and singly charged ions. It may be noted that previous plasma HHG studies were carried out using the time-integrated methods of plasma emission analysis, so it was impossible to define exactly what plasma conditions existed during the propagation of the femtosecond pulse through the plume. Below, we present the time-integrated analysis of the spectral studies of plasma emission from various metal targets used for harmonic generation during propagation of the picosecond radiation through the plasma. These studies were aimed to define the optimal plasma conditions for the efficient HHG in laser plumes and showed that, although, for most plasma plumes, the over-excitation during laser ablation leads to drastic decrease of harmonic generation efficiency; in some cases, one can achieve the conditions for extension of the harmonic cutoff using the picosecond pulses.

The spectral studies of atomic and ionic emission from the laser-produced plasmas were carried out in the visible and near-ultraviolet (UV) ranges (300–700 nm). This spectral region was chosen due to existence of multiple ionic and atomic transitions of the studied plasma species. To create the ablation, a 1064 nm, 38 ps

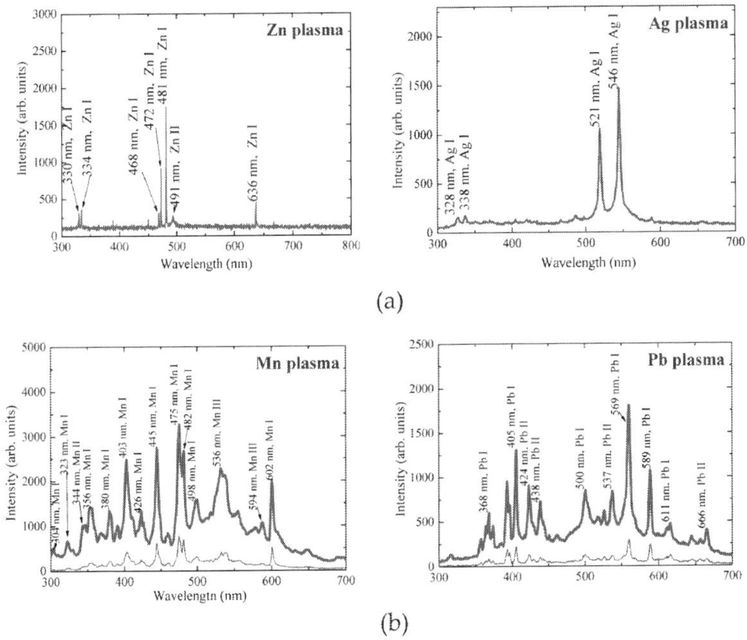

Figure 4.2 (a) Emission spectra from the Zn and Ag plasmas used for harmonic generation. (b) Plasma emission spectra from the Mn and Pb targets at weak (thin lines, $I = 8 \times 10^{10}$ W/cm^2) and strong (thick lines, $I = 2 \times 10^{11}$ W/cm^2) excitation of target surfaces. Reproduced from Ref. 4 with permission from IOP Publishing. All rights reserved.

pulse from Nd:YAG laser was focused on to the metal targets in the vacuum chamber. The spectral characteristics of laser plasma in the visible and UV ranges were analyzed using the fiber spectrometer.

The spectra of plasma emission from the Zn and Ag plasmas are presented in Fig. 4.2a. One can note that these spectra were obtained at the conditions of "optimal plasma" formation from the point of view of best HHG conversion efficiency. For this, we obtained maximum yield of harmonics in the 80–220 nm range and then measured the spectral characteristics of laser plasmas at these conditions. The spectral lines mostly originated from the excited states of neutral and singly charged ions, when the impeding influence of free electrons on the high-order nonlinear optical processes in laser plasmas was insignificant.

The increase of heating pulse intensity on the target surfaces above 10^{11} W/cm^2 led to both the growth of emission intensity for neutral and singly charged ionic species and the appearance of emission lines from the higher charged particles. The appearance of doubly and triply charged ions and large amount of free electrons immediately followed with considerable decrease of HHG conversion efficiency from almost all the plasma samples. The variations of plasma spectra in that case are presented in Fig. 4.2b, in which one can see the changes of intensity of plasma emission from the Pb and Mn targets. Thin and thick curves correspond to the weak and strong excitation, correspondingly. This figure clearly shows the increase of the intensities of Mn III and Pb II lines, which is correlated with increase of the concentration of multiply charged ions in the plasma plume and correspondingly the concentration of free electrons.

Previous studies of HHG from plasma plumes have analyzed the UV emission spectra from the plasma to prove that over-excited and over-ionized plasmas could dramatically decrease the harmonic intensity [6]. In present studies, the restricting features of harmonic generation dynamics during over-excitation of the targets were observed in the case of most of analyzed plasma plumes. The increase of heating pulse intensity from 1×10^{11} to 3×10^{11} W/cm^2 during HHG experiments with these targets led to appearance of strong plasma emission. At these excitation conditions, the harmonic emission from various plasmas overlapped with the plasma emission. The intensity of generated harmonics became considerably less than at 1×10^{11} W/cm^2 excitation. In the case of most of plasmas, this over-excitation led to entire disappearance of harmonic emission. The variations of HHG efficiency with the growth of heating pulse intensity are depicted in Fig. 4.3. In particular, the 11th harmonic generating from Pb plasma started to decrease with the growth of heating pulse energy above 3 mJ (Fig. 4.3a), which corresponds to the intensity of 1×10^{11} W/cm^2. In another case, we observed a decrease of the fifth, seventh, and ninth harmonics generating in Mn plasma at the pulse energies above 1.5 mJ (Fig. 4.3b).

The characteristic pattern of almost all harmonic spectra from metal plasmas was a featureless sharp decrease of conversion efficiency for lowest orders, which followed with a gradual decrease

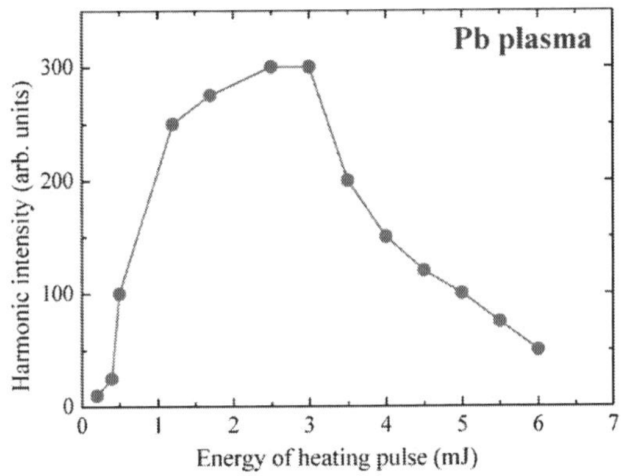

(a)

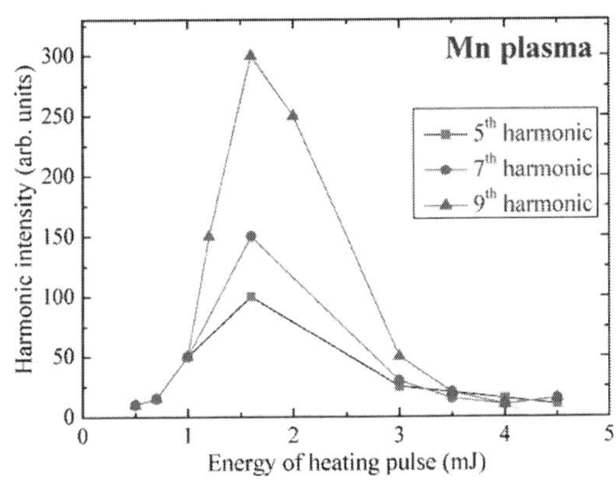

(b)

Figure 4.3 Dependences of the harmonic intensity at different heating pulse energies for the (a) 11th harmonic generating from the Pb plasma, and (b) fifth, seventh, and ninth harmonics generating from the Mn plasma. Reproduced from Ref. 4 with permission from IOP Publishing.

of higher harmonics (above the ninth order) up to the limit of registration range of monochromator (~50 nm, 21st harmonic). Figure 4.4 shows the characteristic HHG spectra obtained from Cr, Cu, and Ti plasmas. Most of these plasmas showed comparable nonlinear optical properties from the point of view of HHG conversion efficiency.

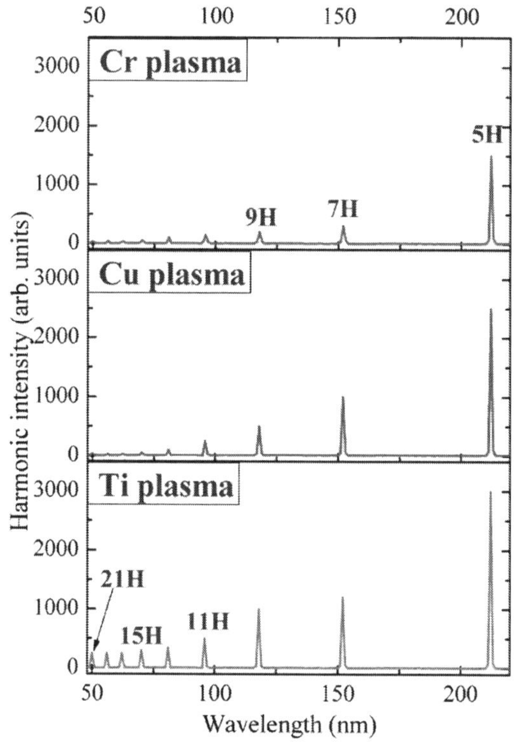

Figure 4.4 High-order harmonic spectra from the Cr, Cu, and Ti plasma plumes. Reproduced from Ref. 4 with permission from IOP Publishing.

The measurements of the absolute value of conversion efficiency of the harmonics generating in plasmas were carried out using the following procedure. At the first step, the fourth harmonic signal was measured by a "monochromator + sodium salicylate + PMT" detection system using known energy of the fourth harmonic of 1064 nm radiation generating in the nonlinear crystals. This allowed the calibration of monochromator at the wavelength of 266 nm. As the quantum yield of sodium salicylate is equal in a broad spectral range between 40 and 350 nm, the calibration of registration system at $\lambda = 266$ nm allowed calculating the conversion efficiency for the higher harmonics. The monochromator allowed observing the harmonics down to the spectral range of 50 nm. The conversion

Table 4.1 Harmonic conversion efficiencies ($\times 10^{-6}$) in various plasma plumes

H	λ	Sn	Zn	Mn	Ti	Cu	W	Au	B	Ta	Ag	Cr	Mo	Al
5	213	100	80	80	70	46	46	40	40	26	20	20	4.2	6
7	152	66	66	20	20	33	40	33	13	10	16	7.1	3.8	2.59
9	118	66	53	10	12	15	38	26	3	3	10	3.8	2.5	1.33
11	97	53	40	6.6	6.6	13	33	20	1.3	1.8	6.6	3.8	2.1	
13	82	26	4	2.1	4.6	2.5	7	1.8	1	0.9	2.6	1.3	0.9	
15	71	16	4	1.75	3.3	0.9	5	1.3		0.9	1.3	0.9		
17	63	6	1.7	1.3	2.1	0.9		1.3		0.9	1	0.9		
19	56	6	1.7	1.3	2.1	0.83		0.9		0.75	1	0.9		
21	51	4	1	1.3	2.1	0.75		0.9		0.75	1	0.9		

Source: Reproduced from Ref. 4 with permission from IOP Publishing.
Note: H, harmonic order; λ, wavelength of harmonic (in nm)

efficiency of plateau harmonics (15th–21st orders) was measured to be in the range of 10^{-6} to 10^{-5}. Table 4.1 summarizes the measured conversion efficiencies for different harmonics in the case of some metal-containing plasma plumes.

Although most of plasmas demonstrated similar properties, some of them allowed the observation of unusual spectral distributions of harmonics. In particular, in some cases, we were not able to observe the generation of harmonics above the 9th–15th orders. Among those plasmas were the Al, B, and Mo plumes (Fig. 4.5). Few other plasmas (In, Nb) showed even more interesting spectra, where one can distinguish the enhancement of some harmonics with regard to lower order ones. Such features resemble those observed in previous plasma HHG experiments using 800 nm, few tens femtosecond pulses in the cases of In, Mn, Cr, Sn, and other plasmas [6]. The commonly accepted explanation of these enhancements was related with the closeness of the specific harmonic orders with the resonance ionic transitions of these media.

The observation of resonance enhancement in the case of both femtosecond and picosecond probe pulses is related with coincidence of harmonic wavelengths and ionic transitions. In the case of femtosecond pulses, this opportunity has more chances to be realized due to broader bandwidth of harmonics. However, the conditions of resonance enhancement could be realized in the

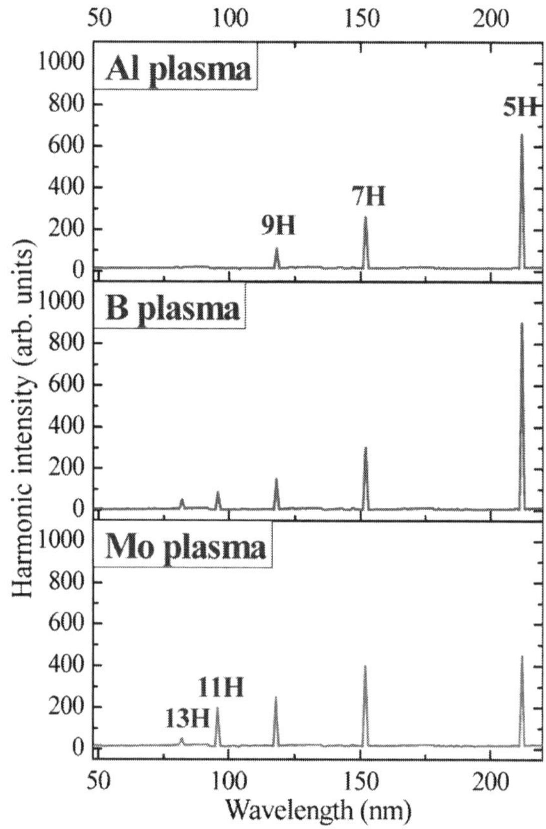

Figure 4.5 Harmonic spectra from the Al, B, and Mo plasmas producing only low-order harmonics. Reproduced from Ref. 4 with permission from IOP Publishing. All rights reserved.

case of picosecond pulses as well once some of ionic transitions occasionally coincide with the wavelengths of harmonics. These observations were repeated several times during analogous studies of the same plasmas (In, Nb), although neither of other plasma plumes demonstrated those features. The propagation effects could not explain these observations, as in that case, we have to observe the enhancement of at least few neighboring harmonics as well. One can note that the resonance enhancement of plasma harmonics in the case of narrowband driving pulses is indeed a rarely observed phenomenon. The application of various plumes showed this enhancement only in a few plasmas, although much more plasmas were reported as suitable for resonance enhancement in the case of femtosecond broadband pulses. These observations show that atomic number does not play a significant role in classification of harmonic properties of the plasma media. The important parameter here could be a second ionization potential of the used species.

A few properties of the generating harmonics from metal-ablated plasmas were analyzed in detail. The delay between heating and driving pulses is crucial for optimization of the HHG. A typical dependence of the harmonic intensity on the delay between pulses is presented in Fig. 4.6a in the cases of Pb and Mn plasmas. One can note that there are different optimal delays for the harmonics generating in the Mn and Pb plasmas (20 ns and 90 ns correspondingly). These observations can be explained by different velocities of ablated particles from these targets, which should reach the area of interaction with the driving pulses at different times. In this connection, the application of heavy target (Pb, Z = 82) should lead to longer delays, at which the bulk amount of ablated material interacts with laser beam, compared with lighter target (Mn, Z = 25). The ratio between optimal delays for these plasma samples approximately coincides with the ratio between the masses of those particles, which should follow from the equal kinetic energies of ablated particles.

The important parameter of plasma HHG is a polarization of probe radiation. Figure 4.6b shows the dependences of the fifth and seventh harmonics generating from Ti plasma at different angles of rotation of the half-wave plate, which caused variation of the conditions of driving radiation from linear (at 0°) to circular (at 45°) polarization. Small deviation from linear polarization led to a considerable decrease of the fifth and seventh harmonic intensity, which is a typical behavior for high-order harmonics. The application of circularly polarized laser pulses led to complete disappearance of harmonic emission.

Harmonic intensity also considerably depended on the distance between the optical axis of driving beam and target surface (Fig. 4.6c). This dependence is induced by the change of plasma concentration above the target. The log-log dependence of harmonic

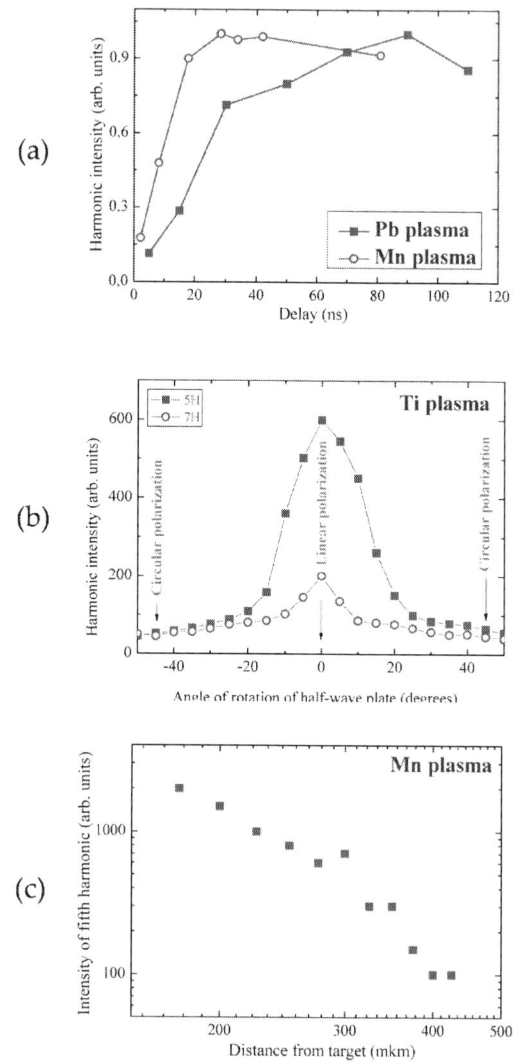

Figure 4.6 (a) Dependences of the harmonic intensity on the delay between heating and driving pulses for the Pb and Mn plasmas. (b) Polarization dependences of fifth and seventh harmonic intensities at different angles of rotation of the half-wave plate in the case of Ti plasma. (c) Harmonic intensity as a function of the distance between the target surface and driving beam axis for the fifth harmonic generating in Mn plasma. Reproduced from Ref. 4 with permission from IOP Publishing.

intensity (I_h) on the distance between the optical axis of driving beam and target surface (x) for the seventh harmonic generation in manganese plasma corresponded to $I_h \sim x^{-3.5}$.

4.2 High-Order Harmonic Generation of Picosecond Laser Radiation in Carbon-Containing Plasmas

The carbon-containing materials such as graphite, fullerenes, and carbon nanotubes have proven that, at optimal ablation conditions, they can be considered as attractive plasma media for the harmonic generation in the 40–80 nm spectral range using the femtosecond laser pulses [7–9]. Propagation effects during the HHG of subpicosecond KrF laser ($\lambda = 248$ nm, $t = 700$ fs) in carbon plasma were studied in [9], and the harmonics up to the 13th order were reported. The studies [7, 8, 10] have shown some peculiarities distinguishing the carbon plasma from other species used for frequency conversion of the femtosecond pulse propagating through the plasma plume. This may be attributed to the specific properties of carbon atoms and ions, or creation of nanoparticles during plasma formation.

Application of picosecond pulses for plasma HHG has some positive preferences compared with femtosecond pulses. As it was underlined in previous section, this preference is related with higher energy density of harmonics at the same level of HHG conversion efficiency. Also, for this purpose, one can use relatively low-cost lasers. The motivation of the studies of the HHG in carbon-containing plasmas using relatively long laser pulses was based on the following reasons: (a) stronger energy conversion efficiency for lower order harmonics, (b) application of the picosecond pulses of relatively high energy, and (c) presence of strong transitions of ionic and neutral carbon in the vicinity of the seventh harmonic of Nd:YAG laser radiation. The goal of these studies was a search of the conditions for generation of energetic coherent picosecond pulses in the range of 80–220 nm using various carbon-containing plasma plumes [11].

The experimental setup of these studies was analogous to the one described in previous section. The targets were made of various carbon-containing materials (graphite, glassy carbon, pyrographite, boron carbide, silicon carbide, and pencil lead). The bulk samples of these materials cut with the sizes of $5 \times 5 \times 2$ mm^3 were used in these studies. A three-coordinate manipulator made it possible to move the target along the z-axis and thus control the zone of the interaction of the probe radiation with the plasma relative to the target plane.

Here, I present some special features of HHG in carbon-containing plasmas as the media for efficient lower order harmonic generation in the VUV range using the 38 ps, 1064 nm driving pulses. At soft ablation conditions, the carbon-containing plasma mainly consisted of excited neutral atoms and singly ionized carbon, which was confirmed by the spectral measurements in near-UV and visible ranges (Fig. 4.7).

Figure 4.8a shows the HHG spectrum obtained in graphite plasma plume. The plateau-like shape of the highest harmonics (above the 13th order) does not exactly represent the approximately equal intensity harmonics. Other carbon-containing plumes showed the same properties (see, e.g., the harmonic spectra from pencil lead and glassy carbon plasmas, Fig. 4.8b,c). It should be noted that the harmonics disappeared after the over-excitation of target surfaces and generation of strong C III and C IV lines in the plasma spectra at high intensities of heating pulse. The maximal conversion efficiency from graphite plasma in the range of plateau (15th–21st harmonics) was measured to be nearly 0.7×10^{-6}.

The important peculiarity of all these studies was the generation of efficient seventh harmonic (Fig. 4.8a–c). Its efficiency considerably (three to seven times) exceeded that of the lower (fifth) order harmonic, contrary to the expected dependence defined by a perturbative theory of lower order harmonics [12]. The enhanced seventh harmonic was a common feature of all used carbon-containing plasma plumes, excluding the boron carbide plasma (Fig. 4.9). In particular, the maximum conversion efficiencies of the fifth and seventh harmonics in the glassy carbon plasma were 10^{-5} and 6×10^{-5}, respectively. The possible reasons of such behavior of the low-order harmonics generating from carbon plasma are discussed below. Note that, in the case of various metal plasmas (Mn, Cu, and so on), this peculiarity (i.e., enhanced 7th harmonic) was not

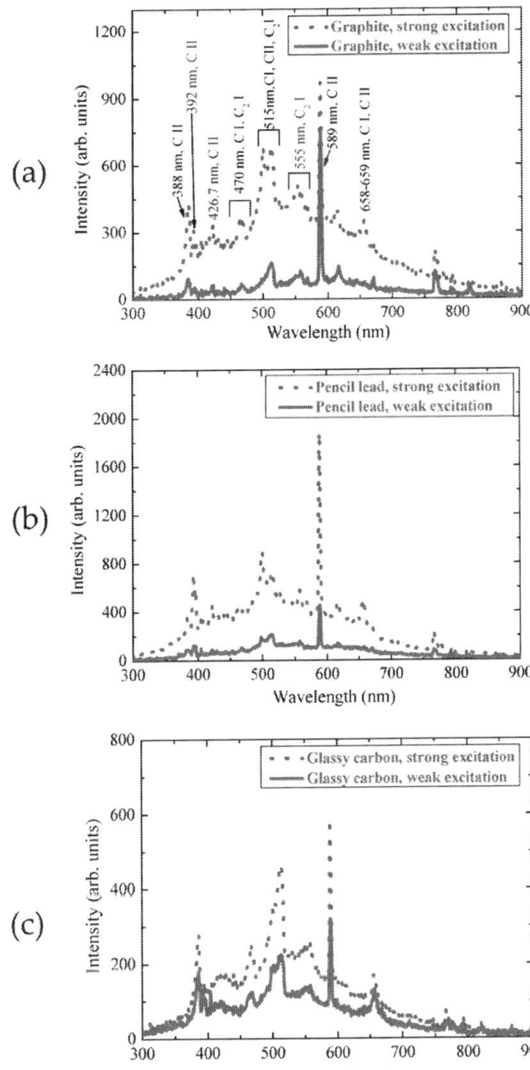

Figure 4.7 Plasma emission spectra from the (a) graphite, (b) pencil lead, and (c) glassy carbon at weak (solid lines, $I = 7 \times 10^{10}$ W/cm^2) and strong (dotted lines, $I = 1.5 \times 10^{11}$ W/cm^2) excitation of target surfaces. Reproduced from Ref. 11 with permission from Optical Society of America.

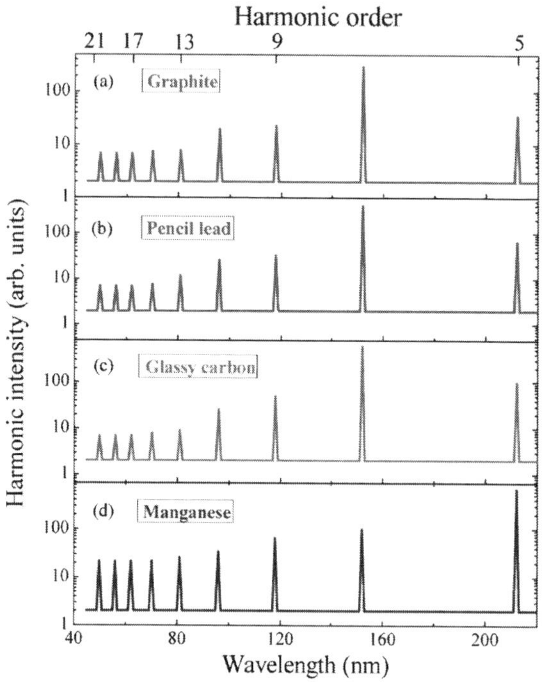

Figure 4.8 Harmonic spectra from the (a) graphite, (b) pencil lead, (c) glassy carbon, and (d) manganese plasmas. Reproduced from Ref. 11 with permission from Optical Society of America.

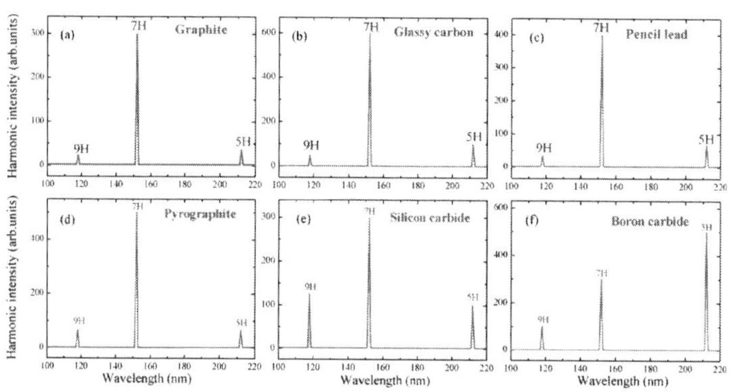

Figure 4.9 Distributions of the fifth, seventh, and ninth harmonic intensities in the cases of different carbon-containing plasmas. (a) Graphite, (b) glassy carbon, (c) pencil lead, (d) pyrographite, (e) silicon carbide, (f) boron carbide. Reproduced from Ref. 11 with permission from Optical Society of America.

observed. The example of the harmonic spectrum from metal (Mn) plasma is presented in Fig. 4.8d, which shows a monotonic decrease of each next order of harmonics over all spectral range up to the 11th order, above which the harmonics showed a plateau-like behavior described by the nonperturbative three-step model of HHG [13].

We analyzed various properties of the seventh and neighboring harmonics. Figure 4.10a shows the dependences of the fifth and seventh harmonics on the angle of rotation of the half-wave plate, which caused variation of the polarization of driving radiation from linear (at 0°) to circular (at 45°). Small deviation from linear polarization led to a considerable decrease of the fifth and seventh harmonic intensities. The application of circularly polarized laser pulses led to the complete disappearance of harmonic emission, as it should be assuming the origin of HHG [13, 14].

The delay between the heating and driving pulses is crucial for optimization of the HHG. A typical dependence of the seventh harmonic intensity on the delay between pulses is presented in Fig. 4.10b in the case of pyrographite plasma. At initial stages of plasma formation and spreading out of the target surface, the concentration of particles (neutrals and singly charged ions) is insufficient, as the particles possessing the velocities in the range of 5×10^3 m/s cannot reach the optical axis of the propagation of driving beam (~50–100 μm above the target surface). The increase of delay allows the appearance of plasma particles along the path of converting pulse, which leads to the growth of HHG efficiency. Further increase of delay leads to saturation of the HHG at nearly 25 ns and gradual decrease of conversion efficiency at longer delays (>40 ns).

There was a similarity in harmonic-delay dependence for all of the samples under investigation. To our opinion, this behavior is caused by the prevailing influence of carbon atoms and ions, as the main emitters of harmonics. Although maintaining the same velocity (~5×10^5 cm/s, in accordance with the estimates based on the thermodynamic approach and a three-dimensional molecular dynamical simulation of laser ablation of graphite using the molecular dynamics code ITAP IMD of the spreading of the carbon particles from the target after ablation [15]), one can calculate the time of their arrival in the area of driving picosecond pulse propagation, which was approximately 100 μm above the

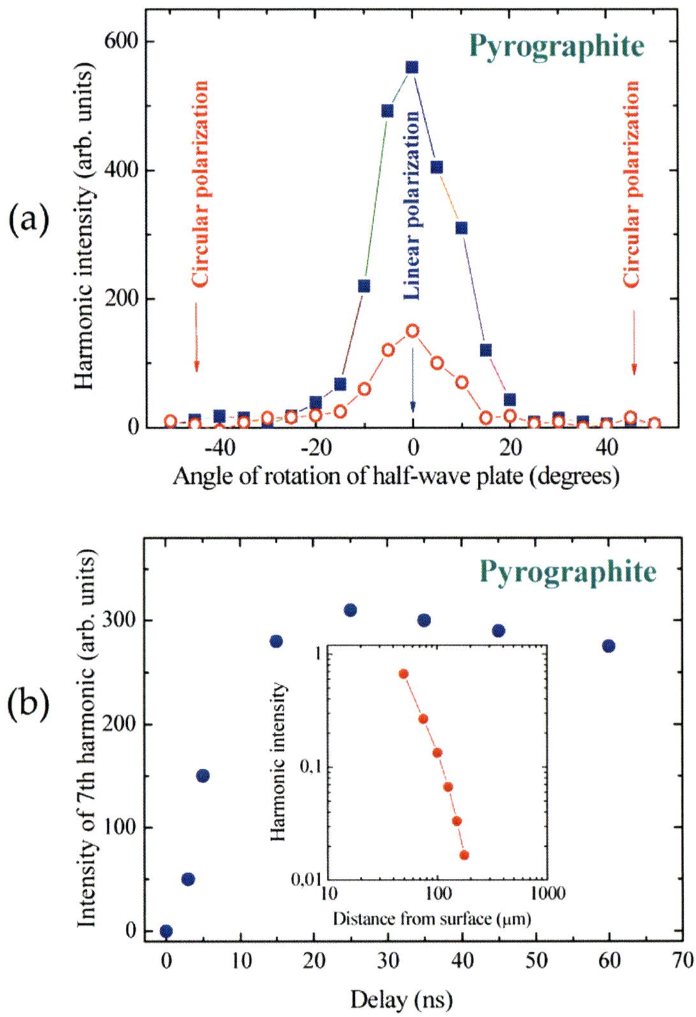

Figure 4.10 (a) Polarization dependences of the fifth (empty circles) and seventh (filled squares) harmonic intensities at different angles of rotation of the half-wave plate. (b) Dependence of the seventh harmonic intensity on the delay between the heating and driving pulses. Inset: Harmonic intensity as a function of the distance between the target surface and driving beam axis for the seventh harmonics. Reproduced from Ref. 11 with permission from Optical Society of America.

target surface, as 20 ns, which is close to the experimental observations. One can note that the delay dependence was similar for any harmonics in the generating spectra from the graphite plasma. Boron carbide showed the same features, as in that case, the main contributors to harmonic yield were the singly charged ions and neutrals as well. This target had the advantage to withstand the laser ablation for a longer period than other targets under study; however, the nonlinear optical properties of B_4C-ablated plume were approximately the same as for graphite and other samples, excluding the absence of single harmonic enhancement in the former case.

Harmonic intensity considerably depended on the distance between the optical axis of driving beam and target surface (see the inset in Fig. 4.10b), due to the change of plasma concentration above the target. The log-log dependence of harmonic intensity (I_h) on the distance between the optical axis of driving beam and target surface (x) for the seventh harmonic generation in pyrographite plasma was $I_h \sim x^{-3}$.

The important parameter for optimization of harmonic emission is the fluence of the heating pulse on the target surface. We measured the dependences of HHG efficiency on the energy of heating pulse, while maintaining the same size of ablation beam on the targets (~500 μm). Figure 4.11 shows these dependences for different harmonics from the pyrographite target. The common feature of these studies was an abrupt decrease of harmonics at irradiation of targets using stronger heating pulses. The reason of these observations has already been discussed in previous chapters and related with the over-excitation of target, which leads to appearance of the abundance of free electrons in the plasma plume. The latter causes a phase mismatch between the waves of driving field and harmonics. This effect is especially important for lower order harmonics.

It has been shown previously that efficient harmonic generation is observed only in the case when the visible and UV plasma emission originated dominantly from the neutral and singly ionized carbon species [16]. The present studies have also confirmed this peculiarity (Fig. 4.7). The broad features near 470, 515, and 555 nm (Fig. 4.7) could be assigned to the bands of excited C_2 molecules.

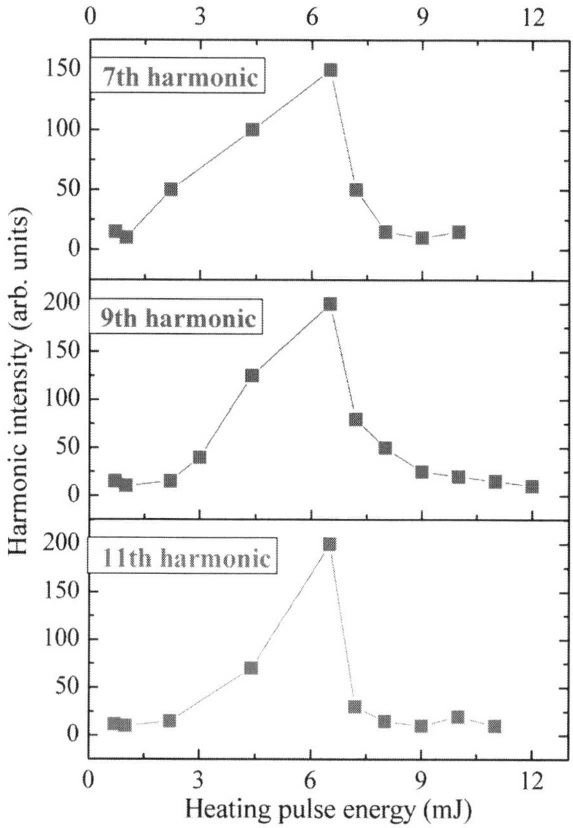

Figure 4.11 Dependences of the harmonic intensity on the heating pulse energy for the seventh (upper curve), ninth (middle curve), and 11th (bottom curve) harmonics generating in pyrographite plasma. Reproduced from Ref. 11 with permission from Optical Society of America.

Other lines in the spectra presented in Fig. 4.7 are attributed to the neutral and singly charged carbon. The main goal of these spectral studies was to analyze what the consistency of carbon plasma was at the conditions of efficient HHG. In the meantime, the analysis of optical spectra in the visible and UV ranges does not inform about the presence of highly ionized species, which can be revealed by collecting the plasma emission in the VUV range. It is worth noting that our observations of harmonic spectra did not show a presence of plasma emission from high-charged particles in the studied spectral range (50–250 nm), which confirms that these experiments were carried out at a soft ablation regime of carbon-containing targets.

The observed difference between the HHG from the plasmas produced on boron carbide and all other carbon-containing targets can be attributed to different influence of the neutral and ionic transitions of B_4C in the vicinity of 152 nm on the harmonic response compared with other species. The tuning of the frequencies of those transitions in the case of B_4C compared with carbon can drastically change the conditions of resonance enhancement for the seventh harmonic. The difference in the nonlinear optical properties of boron carbide and other carbon-containing plasmas also demonstrates that the former molecule probably was presented in the plume mostly without the disintegration during laser ablation; otherwise, the response from singly charged carbon ions could cause the enhancement of seventh harmonic such as in other cases. It means that plasma HHG can be served as a precise tool for the nonlinear spectroscopy analysis of the atomic physics and structure of materials.

Carbon possesses some strong neutral and ionic transitions in the vicinity of the seventh harmonic of Nd:YAG laser radiation. Those are the $2s^22p^2\ ^1\mathrm{D}$ – $2s^22p3d^1\mathrm{D}^0$ (C I, $\lambda = 148.2$ nm), $2s^22p^2\ ^3\mathrm{P}$ – $2s2p^3\ ^3\mathrm{D}^0$ (C I, $\lambda = 156.0$ nm), and $1s^22s^1\ ^2\mathrm{S}$ – $1s^22p^2\mathrm{P}^0$ (C IV, $\lambda = 154.8$ nm) transitions [17]. One can note that, at used experimental conditions, the excitation of triply charged carbon ions is almost impossible, as we excited the targets using the moderate fluences. Probably, the transitions from neutrals ($\lambda = 148.2$ nm and 156.0 nm) are responsible for the observed enhancement of seventh harmonic ($\lambda = 152$ nm). Another option for the explanation of the enhanced seventh harmonic with regard to the fifth order could be the absorption of the latter harmonic in the carbon-containing plasma. However, previous studies of carbon plasma did not show strong absorption lines in the vicinity of fifth harmonic ($\lambda = 213$ nm) [17]. Thus, the most probable reason of observed feature in the vicinity of 152 nm is the resonance-induced enhancement of the seventh harmonic. This enhancement can be explained by the micro and macro-processes influencing HHG.

Below, we present the calculations of the contributions of the enhancement of resonant nonlinear susceptibility and the radiation propagation effect during nonlinear wave interaction in a plasma medium within the framework of perturbation theory, as in our experiments, the maximum amplitude of a laser field (1.7×10^{10} V/m) was nearly 30 times less than intra-atomic field of unexcited carbon atoms (4.7×10^{11} V/m). The nonlinear medium basically consisted of the atoms and singly charged ions of carbon. The carbon atoms can be presented in singlet and triplet states. According to the estimations, which have been carried out using the equations presented in [12, 18] for the observed transitions from unexcited triplet $2s^2\,2p^2\,{}^3\mathrm{P}$ and singlet $2s^2\,2p^2\,{}^1\mathrm{D}$ states of carbon atoms and from unexcited carbon ions, the ratios of the absolute values of seventh and fifth order nonlinearities $r = \left|\chi^{(7)}/\chi^{(5)}\right|$ do not exceed nearly 7×10^{-12} esu. In these calculations, the values of transition frequencies and oscillator strengths were taken from [17]. The estimations have shown that, due to closeness to the resonance conditions, the value of resonant seventh-order nonlinearity is nearly six times higher than non-resonant one. One can note that, for carbon atoms, the frequency shift from seven-photon resonance is about nearly 1600 cm^{-1}.

For estimations of the variations of harmonic signal we used the relation defined for the case when the diffraction length of laser beam R_d is longer than the nonlinear medium length L (see, e.g., [12, 15]). At laser intensity of the order of 10^{13} W cm^{-2} and absence of the generalized phase mismatch influence, the ratio of the intensities of seventh and fifth harmonics $\gamma = \mathrm{I}_7/I_5$ does not exceed 5×10^{-2} at harmonic generation from the atomic and ionic ground states. At the peak laser intensity used (4×10^{13} W/cm^2) this relation should be increased in 16 times. So, at this level of laser intensity used in our experiment the fifth harmonic should be stronger than the seventh harmonic, once we assume the resonant enhancement of the seventh-order nonlinearity alone. Contrary to these estimates, in the experiment, the seventh harmonic was three to seven times stronger than the fifth one at rather broad range of the variations of laser intensity.

It was shown earlier for elliptically polarized laser field that the higher order harmonic can be more effective than the lowest order one [14]. The enhancement of the fifth harmonic (λ = 53.2 nm) of the driving pulse (fourth harmonic of Nd:YAG laser) in helium contrary to third one (λ = 88.7 nm) was observed in [19]. This enhancement of the higher harmonic was explained by the resonant enhancement of the fifth-order nonlinearity responsible for fifth harmonic generation and caused by closeness to five-photon resonance. In the meantime, the estimations have shown that the mechanism of seventh harmonic enhancement could be different.

Let us consider the influence of the phase conditions on the HHG and analyze the propagation effect. The phase mismatches for the fifth and seventh harmonics are positive once we consider the generation of harmonics from the carbon atoms in singlet ground state and also from the carbon ions in ground state. The phase mismatch of the electronic gas is also positive. In that case the total phase mismatch Δk should be positive. In the meantime, during generation of seventh harmonic in the media containing carbon atoms in triplet ground state, the value of phase mismatch is negative. If the medium mostly consists of atomic carbon in triplet state, the addition of nearly 5% of the electronic gas is necessary for the compensation of phase mismatch. In that case the positive value of Δk is increased for the compensation of phase mismatch. The calculations show that, at strong phase mismatch, the dependence of the harmonic intensity on the medium length demonstrates the oscillating features, and the conditions of stronger seventh harmonic compared with the fifth one are possible. In that case the relative increase of seventh harmonic will be caused by a suppression of fifth harmonic intensity. However, at a medium length of 500 μm the relative increase of the seventh harmonic can be achieved at the concentration of the carbon atoms in the triplet ground states exceeding 1.2×10^{19} cm^{-3}.

In experiments, the saturation of harmonic signal with the growth of laser intensity was observed. One of the mechanisms of saturation can be the phase shift caused by the influence of Kerr nonlinearities. The estimations show that, at laser intensity of 10^{13} W/cm^2 and medium concentration of 10^{18} cm^{-3}, the absolute value of nonlinear phase mismatch was about 50 cm^{-1} for carbon atoms during the excitation from the unexcited state. The nonlinear phase mismatch for the fifth harmonic is positive, whereas that for the seventh harmonic is negative. So, the nonlinear phase mismatch of seventh harmonic compensates the phase mismatch of free electrons. In that case, at the concentrations of atomic carbon in the ground state of 2×10^{18} cm^{-3} and the electronic gas more than 6×10^{17} cm^{-3}, the seventh harmonic exceed the fifth one at the laser intensity less than 10^{13} W/cm^2. The nonlinear phase shift also saturates the harmonic intensity, which stabilized the ratio of the seventh and fifth harmonics. Thus, the influence of Kerr nonlinearities can explain the relative enhancement of the seventh harmonic and its stabilization over rather broad range of laser intensities.

The estimations, which were carried out for the HHG from the exited state $2s^22p3s^3\mathrm{P}^0$ of carbon atoms, have shown that the presence of a small part of exited particles ($\sim 10^{-3}$) is capable to considerably change the total linear phase mismatch and nonlinear susceptibilities responsible for harmonic generation and self-action effects. The analysis of this exited state is of special interest, as in that case rather strong single-photon resonance with the laser radiation can be observed (e.g., transition $2s^22p3s^3\mathrm{P}^0 - 2s^22p3p^3\mathrm{D}$, frequency shift $\sim$37 cm^{-1}).

Regarding the main finding of resonance enhancement of single harmonic in the carbon plasma, one can consider it not as just a clear evidence of considerable enhancement of single harmonic in the spectral range where no such phenomenon was reported so far, but rather as a tool for the spectroscopic studies of various atomic and ionic transitions possessing different oscillator strengths.

4.3 Resonance Enhancement of Harmonic Generation of 1064 nm Picosecond Radiation in Lead Plasma

Some metal plasmas, such as indium, manganese, and tin [20, 21], have proven that, at appropriate experimental conditions, they can be considered as excellent plasma media for the enhanced harmonic generation in the 80 to 40 nm spectral range using ultrashort laser pulses. Moreover, some harmonic orders (13th harmonic from In plasma and 17th harmonic from Sn plasma in the case of 800 nm, femtosecond radiation) demonstrated the enhancement attributed to the influence of the ionic transitions. The mechanisms of such effect were discussed in Chapter 2.

A study of the harmonic generation of relatively long (picosecond) pulses of Nd:YAG laser in the plasmas of metals was described in previous section. An analysis of low-order harmonic generation in such plasmas produced by a heating pulse on the surface of a solid target allowed several recommendations regarding further advancement toward the shorter wavelengths. Recent studies of these plasma plumes showed the observation of unusual distribution of harmonic spectra from Pb plasma, when the 11th harmonic was stronger than lower-order harmonics [22].

The enhancement of 11th harmonic from Pb plasma in the case of 1064 nm picosecond driving pulses resembles above-mentioned peculiarities of resonance-induced enhancement of harmonics. A thorough analysis of lead harmonics would allow the understanding of the physical mechanisms leading to such enhancement. In this section, I analyze the results of studies of the HHG in the lead plasma. We compare those enhanced harmonics in the cases of laser ablation of pure Pb and its alloy. We also analyze the resonance-induced enhancement of the 11th harmonic of the picosecond radiation of Nd:YAG laser from this plasma in the presence of different gases.

The experimental setup was described in details in Section 4.1. The targets were made of various metals such as lead, tin, copper, and zinc. For comparison, as target material, we also used the alloys of these metals. Various gases were inserted inside the vacuum chamber containing plasma plumes to analyze the influence of absorption and dispersion properties of these gases on the variations of the plasma harmonic spectra.

We present some specific features of lead and alloy plasmas as the sources of HHG using picosecond laser pulses. The goal of these studies was to define whether the components of Pb-containing plasma influence the resonance-enhanced harmonic generation efficiency. Changing the composition of plasma from pure lead was accomplished using an alloy of this material, which consisted of tin and lead (at a volume ratio of ~5:3). We also compared plasma harmonic spectra from pure lead and alloy with those from pure tin. Another alloy target was brass, which consisted of copper and zinc (65%/35%). In addition, we studied the variations of harmonic

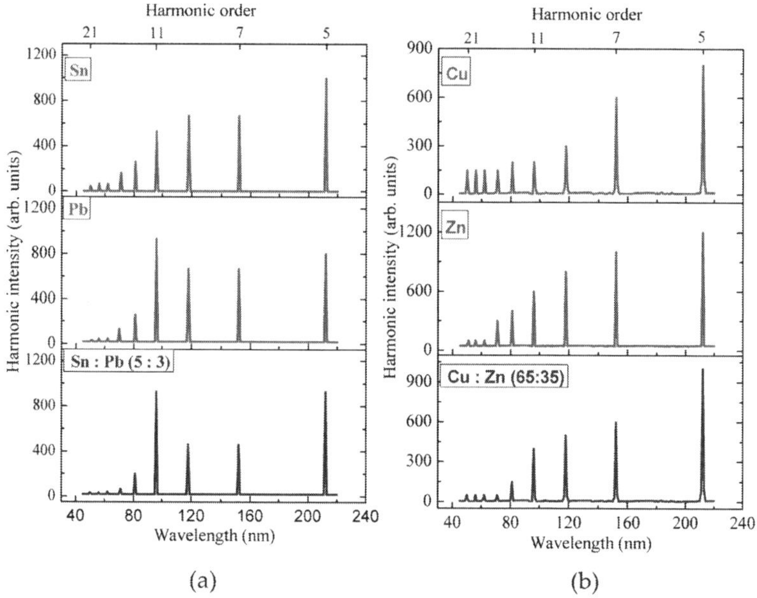

Figure 4.12 Harmonic spectra from the (a) Sn, Pb, and Sn:Pb (5:3) alloy plasmas, and (b) Cu, Zn, and Cu:Zn (65%/35%) alloy plasmas. Reproduced from Ref. 22 with permission from Optical Society of America.

spectra from lead plasma while adding different gases in the vacuum chamber, thus changing the components inside the plasma plume.

Figure 4.12a shows the HHG spectra obtained in the Sn, Pb, and Sn/Pb alloy plasma plumes. The harmonics from tin plasma (Fig. 4.12a, upper panel) showed a gradual decrease of each next harmonic's intensity, which is a common feature in the case of most of plasma harmonics experiments [16]. Analogous harmonic spectra were observed in the cases of Cu, Zn, and brass plasmas (Fig. 4.12b). The conversion efficiency in the range of 15th–21st harmonics was measured to be nearly 1×10^{-6}.

In the meantime, lead plasma demonstrated the enhanced 11th harmonic (Fig. 4.12a, middle panel), which was stronger than lower orders. This peculiarity of lead harmonics was maintained at variations of different conditions of experiments (confocal parameter, plasma length, and so on). The harmonics from mixture of lead and tin plasmas comprised both harmonic spectra, while maintaining the enhanced 11th harmonic (Fig. 4.12a, bottom panel).

We analyzed the harmonics from lead plasma at variable experimental conditions. The delay between heating and driving pulses is crucial for optimization of the HHG. A typical dependence of 11th harmonic intensity on the delay between pulses is presented in Fig. 4.13a. At the initial stages of plasma formation and spreading out of the target surface, the concentration of particles (neutrals and singly charged ions) is insufficient, as the particles possessing the velocities in the range of 5×10^5 cm/s do not reach the optical axis of probe beam propagation ($\sim$100–150 μm above the target surface). The increase of delay allows the appearance of plasma particles along the path of converting pulse, which leads to the growth of HHG efficiency. Further increase of delay leads to saturation of the HHG at nearly 70 ns and gradual decrease of conversion efficiency at longer delays (>110 ns). We also investigated the influence of the distance between the target and the optical axis of propagation of the fundamental radiation (see the inset in Fig. 4.13a). This distance was varied by a manipulator, which controlled the position of the target relative to the waist of the converting picosecond radiation.

The common feature of these studies was an abrupt decrease of harmonics at irradiation of targets above the appropriate level. The reason of these observations is obvious and related with the over-excitation of plasma, which leads to the appearance of abundance of free electrons. Figure 4.13b shows the influence of the energy of heating pulse on the intensity of 11th harmonic from the lead plasma. Harmonic intensity increased up to the heating energy of 3 mJ and further growth in energy led to a decrease of the intensity of harmonics.

The analysis of influence of fundamental intensity on efficiency of 11th harmonic generation in lead plasma was carried out. The dependence of harmonic signal on fundamental intensity is shown in Fig. 4.14. The slope of this dependence was equal to 1.3 (dashed lines). With growth of laser intensity, the decrease of power dependence occurred practically monotonously without presence of obvious oscillations. As frequency shift from 11th photon resonance between ground and exited states $6s^26p\ ^2P_{1/2}$ – $6s^28d\ ^2D_{3/2}$ was nearly 4.5 cm^{-1} only, the analysis of given behavior was carried out

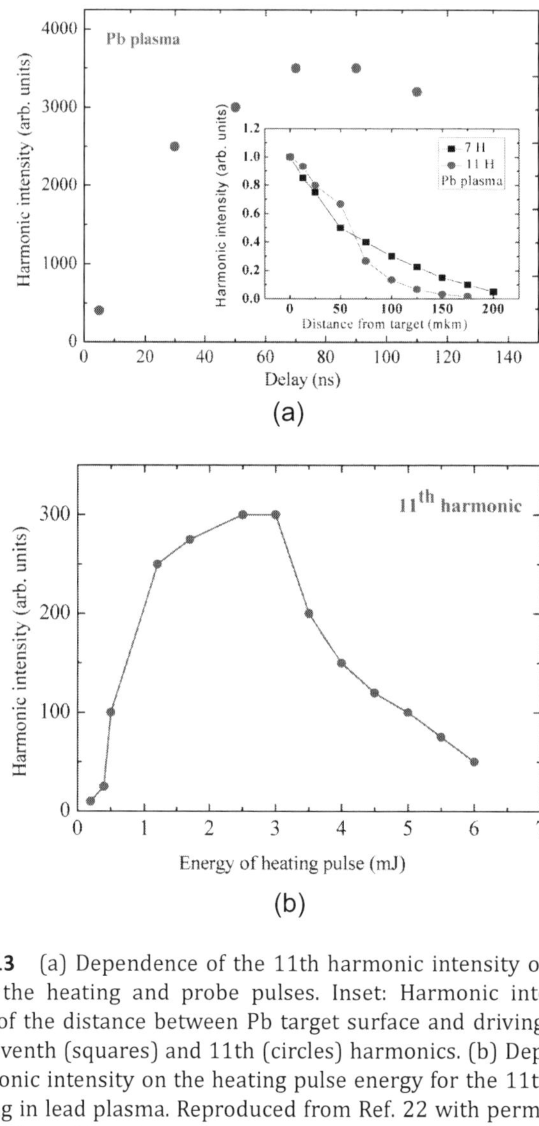

Figure 4.13 (a) Dependence of the 11th harmonic intensity on the delay between the heating and probe pulses. Inset: Harmonic intensity as a function of the distance between Pb target surface and driving beam axis for the seventh (squares) and 11th (circles) harmonics. (b) Dependence of the harmonic intensity on the heating pulse energy for the 11th harmonic generating in lead plasma. Reproduced from Ref. 22 with permission from Optical Society of America.

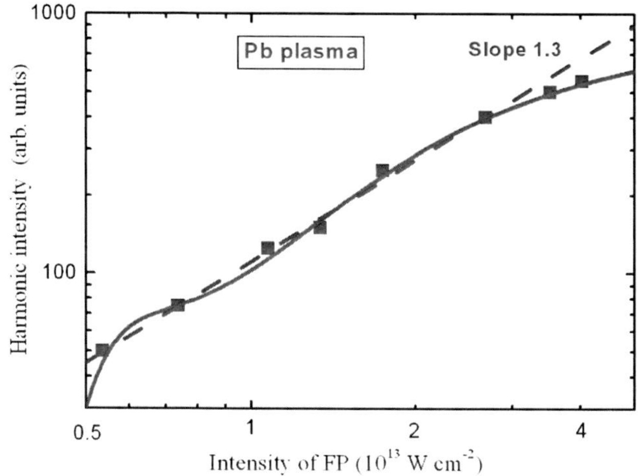

Figure 4.14 Comparison of the experimental (filled squares) and calculated (solid curve) dependences of the 11th harmonic on the driving pulse intensity. Reproduced from Ref. 22 with permission from Optical Society of America.

in resonant approach [18]. In Fig. 4.14, the calculated dependence obtained at fitting of ac Stark shift, relaxation time, and generalized matrix elements is presented by a solid line. From this figure, one can see that the influence of resonance well enough describes the process of 11th harmonics generation in lead plasma.

The insertion of gases in the vacuum chamber containing targets led to variations of harmonic spectra compared with the case of pure lead plasma. We used four noble gases (He, Ar, Kr, Xe) possessing different dispersion, as well as different absorption, in the region of interest (50–250 nm). Figure 4.15a,b shows the Pb harmonic spectra in the presence of light (He) and heavy (Xe) gases. The former gas possesses weak absorption until 65 nm and thus its influence on the variations of harmonic spectra could be attributed to the optical dispersion properties of gas changing the phase matching conditions for resonant and non-resonant harmonics. We used the He pressures up to 100 torr above which the optical breakdown of the gas (without ignition of the metal plasma) was observed. One can see the abrupt decrease of 11th harmonic with the increase of He pressure. Meanwhile, some harmonics, in particular the ninth one,

(a)

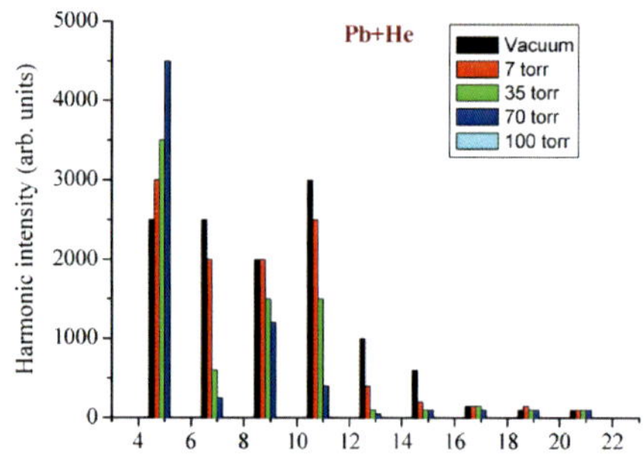

(b)

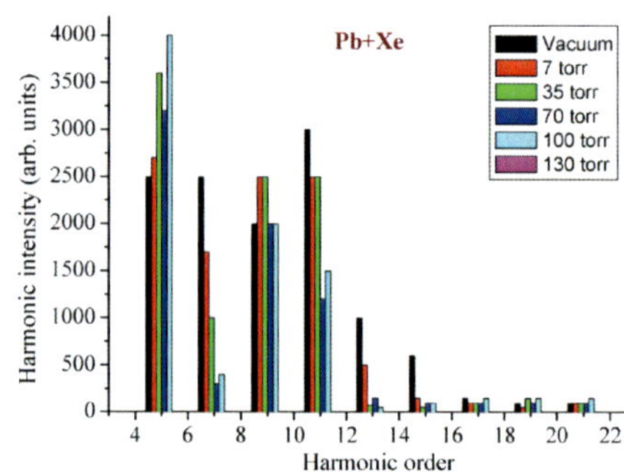

(c)

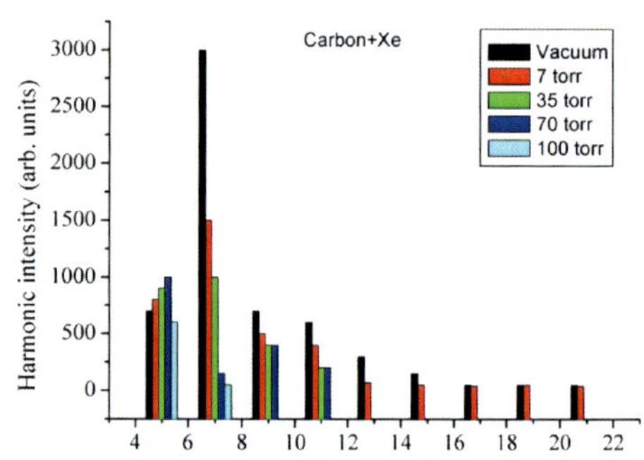

Figure 4.15 Variations of plasma harmonic spectra at different pressures of gases. (a) Pb plasma, He gas, (b) Pb plasma, Xe gas, (c) carbon plasma, Xe gas. Reproduced from Ref. 22 with permission from Optical Society of America.

showed less decrease, and the fifth harmonic became even stronger than the plasma formation at vacuum conditions. Approximately same features were observed in the case of insertion of xenon, while a decrease of 11th harmonic was not so pronounced.

We compared the influence of gases in the harmonic spectra from the plasma media showing the enhancement of specific orders. Recently, the studies of the plasma harmonics from graphite ablation showed a strong seventh harmonic of Nd:YAG laser radiation (152 nm) compared with the fifth harmonic ([11], also see previous section). The enhancement of this harmonic was attributed to the closeness with ionic transition of carbon and improvement of phase matching conditions for this particular case. Here, in Fig. 4.15c, we present the spectra of carbon harmonics at different pressures of xenon. One can admit that a decrease of strong seventh harmonic with the growth of gas pressure (although the neighboring harmonics did not follow this abrupt change of intensity) resembles the one shown in the case of lead plasma (Fig. 4.15b). The comparable variations of Pb and C harmonic spectra may lead to the following conclusions. The enhancement of harmonics at vacuum conditions in both these cases originated from the better conditions of the phases of fundamental and harmonic waves (11th and seventh harmonics from the lead and carbon plasmas, respectively). The addition of the medium possessing positive dispersion worsens the phase matching between these waves and correspondingly decreases the intensities of those harmonics. The role of the gas on other harmonics is also defined by the positive addition to the refractive index of the plasma-gas mixture. However, in that case, the dispersion properties of gas play less decisive role compared with the resonance-enhanced harmonics.

Overall, the insertion of gas leads to the change of the influence of propagation effect on harmonic generation. The combination of microprocesses (i.e., closeness of harmonic wavelength and ionic transition, absorption processes) and macroprocesses (propagation effect comprising the joint influence of Gouy phase, dispersion of plasma near resonant transitions, and dispersion of neutral gas) results in the observed variations of harmonic spectra from the plasma plumes at different pressures of gas. Notice that the concentration of lead particles in the area of interaction with driving pulse ($\sim 2 \times 10^{17}$ cm^{-3}) becomes comparable with the one of gases [(1–10) $\times$ 10^{17} cm^{-3}] inside the plasma volume. One can remind that, at the early stages of gas harmonic studies, the inclusion of additional gaseous component allowed the enhancement (or decrease) of generating harmonics in the gas mixtures in the vacuum ultraviolet range [23–26].

It is seen from the results of plasma mixtures (Fig. 4.12) that the role of nonresonant plasma components (e.g., tin in the case of ablating Sn–Pb alloy) was insignificant and did not lead to the considerable variation of the relative harmonics compared with pure Pb plasma plume. This observation points out the insignificant influence of dispersion properties of additional plasma component on the overall relations between the phases of 11th harmonic and 1064 nm waves.

References

1. M. Ferray, A. L'Huillier, X. F. Li, L. A. Lompré, G. Mainfray, G. Manus, *J. Phys. B: At. Mol. Opt. Phys.* 21, L31 (19988).
2. L. A. Lompré, A. L'Huillier, M. Ferray, P. Monot, G. Mainfray, G. Manus, *J. Opt. Soc. Am. B* 7, 754 (1990).
3. A. McPherson, G. Ginson, H. Jara, N. Johann, I. A. McIntyre, K. Boyer, C. K. Rhodes, *J. Opt. Soc. Am. B* 4, 595 (1987).
4. R. A. Ganeev, G. S. Boltaev, N. K. Satlikov, T. Usmanov, *J. Opt.* 14, 095202 (2012).
5. L. B. Elouga Bom, J.-C. Kieffer, R. A. Ganeev, M. Suzuki, H. Kuroda, T. Ozaki, *Phys. Rev. A* 75, 033804 (2007).
6. R. A. Ganeev, *J. Phys. B: At. Mol. Opt. Phys.* 40, R213 (2007).
7. L. B. Elouga Bom, Y. Pertot, V. R. Bhardwaj, T. Ozaki, *Opt. Express* 19, 3077 (2011).
8. Y. Pertot, L. B. Elouga Bom, V. R. Bhardwaj, T. Ozaki, *Appl. Phys. Lett.* 98, 101104 (2011).
9. W. Theobald, C. Wülker, F. R. Schäfer, B. N. Chichkov, *Opt. Commun.* 120, 177 (1995).
10. R. A. Ganeev, *Phys. Uspekhi* 52, 55 (2009).
11. R. A. Ganeev, G. S. Boltaev, N. K. Satlikov, I. A. Kulagin, T. Usmanov, *J. Opt. Soc. Am. B* 29, 3286 (2012).
12. J. F. Reintjes, *Nonlinear Optical Parametric Processes in Liquids and Gases*, Academic Press, Orlando (1984).
13. P. B. Corkum, *Phys. Rev. Lett.* 71, 1994 (1993).
14. N. L. Manakov, V. D. Ovsyannikov, *Sov. Phys. JETP* 52, 895 (1980).

15. R. A. Ganeev, I. A. Kulagin, M. Suzuki, M. Baba, H. Kuroda, *Opt. Commun.* 249, 569 (2005).
16. R. A. Ganeev, *J. Modern Opt.* 59, 409 (2012).
17. NIST Atomic Spectra Database (ver. 4.1.0), http://physics.nist.gov/asd
18. I. A. Kulagin, T. Usmanov, *Quantum Electron.* 30, 520 (2000).
19. J. Reintjes, C. Y. She, *Opt. Commun.*, 27, 469 (1978).
20. R. A. Ganeev, M. Suzuki, T. Ozaki, M. Baba, H. Kuroda, *Opt. Lett.* 31, 1699 (2006).
21. M. Suzuki, M. Baba, R A Ganeev, H. Kuroda, T. Ozaki, *Opt. Lett.* 31, 3306 (2006).
22. R. A. Ganeev, G. S. Boltaev, I. A. Kulagin, N. K. Satlikov, R. I. Tugushev, T. Usmanov, *J. Opt. Soc. Am. B* 31, 481 (2014).
23. G. C. Bjorklund, *IEEE J. Quantum Electron* 11, 287 (1975).
24. R. Hilbig, A. Lago, R. Wallenstein, *Opt. Commun.* 49, 297 (1984).
25. R. A. Ganeev, V. V. Gorbushin, I. A. Kulagin, T. Usmanov, *Appl. Phys. B* 41, 69 (1986).
26. R. A. Ganeev, T. Usmanov, *J. Opt. A* 2, 550 (2000).

Chapter 5

Plasma HHG Using Femtosecond Pulses

Coherent short wavelength radiation is of increasing importance for a broad variety of basic and applied research in various fields of physical, chemical, and life sciences. Among them, femtosecond time-resolved coherent diffractive imaging and photo-induced processes at surfaces and nanoparticles, as well as lithography, plasma diagnostics, and materials science processing and diagnostics are of foremost interest. The HHG from femtosecond laser pulses allows producing coherent radiation in the XUV spectral range. Table-top lasers render these processes possible with the prospect of widespread scientific applications.

To promote the use of XUV radiation, it seems therefore appropriate to advance laboratory-scale sources to a higher application level. Many interesting experiments can be performed by HHG on the basis of laboratory-scale femtosecond lasers. These sources easily cover the spectral range between 10 and 100 eV photon energy of harmonics, and with few-cycles laser systems even up to several 100 eV. For practical applications of high-order harmonic sources, higher conversion efficiency and thus an increase in the photon flux and also of the maximum photon energy of the harmonic radiation would be beneficial. HHG itself can be used as a spectroscopic tool for analysis of the optical, nonlinear

Plasma Harmonics
Rashid A. Ganeev

ISBN 978-981-4463-30-0 (Hardcover), 978-981-4463-31-7 (eBook)
www.panstanford.com

optical, and structural properties of the emitters of harmonic generation currently comprising a few noble gases. The generation of high harmonics in laser-produced plasmas from various solid-state targets, being for this purpose a relatively new and largely unexplored medium, promises to yield these advances.

The reviews on plasma harmonics studies were mostly concentrated on the discussion of specific topics such as the application of the nanoparticle-containing plasmas for the HHG [1], resonance-induced enhancement of harmonics [2], and applications of fullerenes as the attractive media for harmonic generation [3], contrary to the two first topical reviews published half decade ago [4, 5], where the whole range of plasma HHG studies was presented. Recent reviews [6–8] follow this tendency.

In this Chapter, I discuss the realization of new ideas emerged during last few years, which further improved the HHG efficiency through harmonic generation of femtosecond radiation in specially prepared plasmas and allowed the spectral and structural studies of matter through the plasma harmonic spectroscopy. I also present the current status of plasma HHG studies and show new trends and perspectives in the developments of this filed.

5.1 Current Status of Plasma HHG Studies

Plasma HHG has become considerably mature during last few years and continues to attract the growing attention of various laboratories worldwide. Currently, the studies of plasma HHG are carried out in Japan, Canada, India, Uzbekistan, the United States, Ireland, Germany, Korea, Spain, Germany, and United Kingdom. In the following sections, the most recent developments and some fresh approaches, experimental schemes and ideas are described, which could considerably push this field toward the dramatic improvement of the output characteristics of harmonics and better understanding of the material properties through the plasma HHG spectroscopy.

The intense HHG from plasma that is created from different carbon targets was demonstrated recently in [9] using 10 Hz pulse repetition rate laser. A high-order harmonic energy in the multimicrojoule range for each harmonic order from the 11th to the 17th harmonic was obtained. It was concluded, by analyzing the target morphology and the plasma composition, that the intense harmonics from the bulk carbon targets originate from the nanoparticles produced during ablation of the carbon-contained target. It was shown in previous work that nanoparticles and films of C_{60} would generate harmonics that are more intense than those obtained from solid targets [10]. The disadvantage of using nanoparticle and film targets is the instability of the harmonics, which considerably varies from shot to shot, and even disappears after a few laser shots if the target is not moved. On the contrary, in [9], it was found that carbon bulk targets can generate harmonics, comparable to those from the nanoparticle or C_{60} plasma targets.

The scanning electron microscope image of plasma debris from a carbon target revealed that the plasma plumes contain nanoparticles with sizes varying between 100 nm and 300 nm. It was therefore suggested that, during the interaction of the heating pulse with the carbon target, nanoparticles are formed in the plasma and are then pumped by the fundamental pulse to induce the generation of harmonics. The harmonic intensity using bulk carbon target remained stable for several minutes, even without moving the target position. By creating the plasma during 5 min on the same place of the solid carbon target the harmonic intensity does not decrease more than 10%, while the one from nanoparticles decreased more than 90% after a few seconds. It was also noted that, unlike most other solid targets, the highest harmonic obtained with a carbon target does not exceed 21st order. According to the cut-off law defined by the three-step model, it was suggested that these harmonics are generated by neutral atoms, rather than ions as in the case of other solid targets.

Further developments of both the stability of carbon-containing plasma harmonics and their enhanced yield were reported in [11]. The importance of these parameters is defined by the applicability of converted radiation for various needs. Many efforts have been dedicated to the improvement of these characteristics during the long history of harmonic generation in gases. The multi-microjoule harmonics have been generated by energy scaling of gas HHG under highly optimized conditions [12, 13], which, however, have basically pushed gas HHG to its limits. Therefore, there is an urgent need to search for methods to generate even more intense harmonics and attosecond pulses. For these purposes, gas clusters [14] and plasma produced from nanoparticle targets [10, 15] can be used to increase the intensity of harmonics. In these cases, microjoule intense harmonics have been demonstrated in the range of 50–90 nm.

However, as was mentioned above, nanoparticle targets have the problem of a rapid decrease in the HHG intensity with consecutive shots, which prevents them from being used in applications, especially in the case of high pulse repetition rate lasers. These exotic targets are also not always available in abundance. A new approach has been reported recently, which showed that highly efficient and stable high-order harmonics could be generated from a target that is readily available in the household, pencil lead of a mechanical pencil [11]. The measurements of the harmonic energy generated from plasma produced from pencil lead and the comparison with harmonics produced from C_{60} particles, which has proven to be one of most efficient media for plasma HHG, showed the advantages of harmonics from the former medium. The important advantage of using pencil lead target is the shot-to-shot stability of the harmonic intensity over a sufficiently large number of shots.

To understand the uniqueness of the pencil lead plasma, the researchers analyzed the ablated material debris deposited on silicon substrates that were placed close to the ablation plume using a SEM. Those SEM images revealed that the plasma created from a pencil lead target contains nanoparticles whose mean size was in the range of 200 nm. They therefore suspected that due to the ablation of the pencil lead target by the heating pulse, the nanoparticles are formed on the target surface and remained in the plasma, which in turn led to generation of the intense harmonics.

From the experimental observations of stronger harmonics than in case of fullerene plasma and the morphology of plasma debris, the authors of [9, 11] inferred that the origin of the extremely strong harmonics from pencil lead and carbon plasmas is similar to those described for nanoparticle targets. The presence of nanoparticles in the plasma deposition and low cut-off suggest that neutral atoms of nanoparticles are the main source of intense harmonics from the pencil plasma. An explanation for intense harmonic generation from nanoparticles could be the higher concentration of neutral atoms due to the presence of nanoparticles. Unlike single atoms and ions, whose density quickly decreases due to plasma expansion, the nanoparticles retain densities that are close to its solid state. Combined with the higher harmonic efficiency of neutral atoms compared with their ions, the neutral atoms within the nanoparticle could generate high-order harmonics efficiently. The authors of [9, 11] estimated a conversion efficiency of nearly 10^{-4} for the harmonics in the plateau range.

The important issue of HHG from plasmas is related to the characteristics of generated harmonics. Whereas the conversion efficiency issue has been taken seriously during recent developments of this technique, which led to a considerable enhancement of harmonic pulse energy, no temporal characterization of plasma harmonic pulses has been performed up to recent time. This is a crucial element for applications of a new source of coherent XUV radiation. It should not be taken for granted that this harmonic emission has a nice attosecond structure. Indeed, the generation in plasma induces many sources of distortion: the higher electron densities and gradients will affect the generation of harmonics through phase mismatching and may result in distortion of both the harmonic spatial phase-front and spectral phase. Furthermore, the temporal characterization itself raises problems, such as the probe beam distortions, target deterioration and instability of harmonic intensity.

The first measurements of the attosecond emission generated from the under-dense plasma produced on a solid target were reported by Elouga Bom et al. [9]. They generated high-order harmonics of a femtosecond Ti:sapphire laser focused in a weakly ionized chromium plasma. The characterization of the plasma attosecond emission was performed using the "Reconstruction of Attosecond Beating by Interference of Two-photon Transitions" (RABITT) technique [16]. The measurement of the harmonic spectral amplitude and phase allows for direct access to the attosecond structure through a Fourier transform. The amplitude of each harmonic is easily given by the amplitude of the main photoelectron lines corrected for the ionization cross-section.

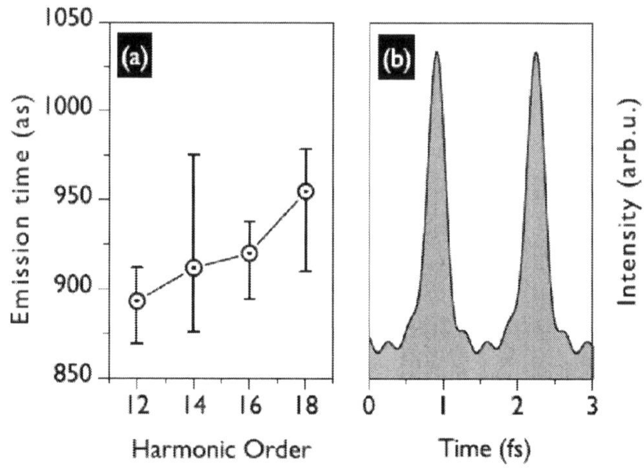

Figure 5.1 (a) Emission times, and (b) temporal profile corresponding to the 11th–19th harmonic orders. Reproduced from Ref. 9 with permission from Optical Society of America.

The relative phase between neighboring harmonic orders is accessed through two-photon XUV + IR ionization of the target gas. When the IR beam is superimposed on the XUV beam in argon gas, sidebands appear in the photoelectron spectrum between the main lines. They correspond to two-photon transitions: absorption of a harmonic photon, $q\omega_0$, accompanied by either absorption or stimulated emission of a laser photon, ω_0. As two coherent quantum paths lead to the same sideband, interferences occur, which result in an oscillation of the sideband amplitude as the delay, τ, between the IR and harmonic field is scanned with sub-IR laser-cycle resolution. The phase of the oscillation is the phase difference between the two interfering channels. The phase difference, ϕ_q–ϕ_{q+2}, between two consecutive harmonics can then be extracted, readily giving the group delay, also called emission time. From the phases, ϕ_q, obtained by integrating the emission times and the amplitudes, A_q, of the harmonic orders, one can reconstruct the temporal intensity profile.

The result for the measured five harmonic orders $q = 11$ to 19 is shown in Fig. 5.1. The reconstructed temporal profile of the harmonic emission forms an attosecond pulse train, with each pulse of 300 as duration (full width at half maximum). Assuming all five harmonics to be in phase, one can obtain the shortest pulses possible, that is, the Fourier-transform limited pulses. The corresponding duration is $\tau = 285$ as. The measured duration of 300 as is thus only 1.05 times the Fourier transform limited duration.

In the standard scheme of gas HHG an ultrafast laser pump beam at intensities above 10^{14} W/cm^2 is focused into a gas jet, generating high harmonics. The yield of such schemes is inevitably limited by dispersion in the medium. Across a distance equal to the coherence length a phase mismatch of π grows and causes destructive interference between the pump and high harmonic beams. This process is one of the major limitations on the conversion efficiency of HHG. Quasi-phase matching (QPM) is a well-known approach for resolving this phase mismatch problem [17]. In QPM, the medium is modulated with a coherence length period so that the pump phase or harmonic emission is changed to prevent the destructive interference caused by the phase mismatch. For HHG in the XUV range and beyond, dispersion in the medium can be mostly attributed to free electrons generated using the laser ionization of the medium. Under this assumption the coherence length (at 0.8 μm wavelength) is given (in meters) by $L_c \propto 10^{15}/qN_e$ where N_e is the free electron density (per cubic centimeter) and q is the harmonic number. Previously, QPM was realized by using multiple gas jets whose pressure and separation were properly controlled [18]. However, the realization of this technique is limited by geometrical constraints on the number and minimal separation of the jets.

It was proposed that the same procedure could be carried out for plasma HHG [19] using a simple method for fabricating numerous plasma jets tailored for the HHG, relieving technical restraints on the dimensions of the jets and their periodicity. In this scheme, the jets are produced by ablation of a microlithographic periodic stripe pattern (Fig. 5.2). Cylindrical plasma jets formed by ablation extend the lithographic pattern into the space above the target, creating a row of narrow plasma jets of different material composition. The efficiency of HHG in plasma has been demonstrated to vary considerably with the atomic composition [20], and the periodic change in this efficiency enables QPM-HHG.

The results of [19] demonstrate a simple method for generation of periodic plasma structures by ablating a lithographic pattern. By

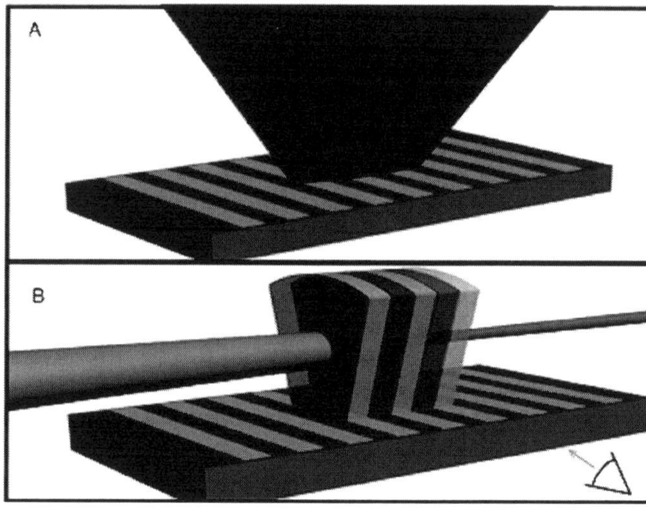

Figure 5.2 System schematics: (a) the lithographic pattern hit by a relatively low-intensity laser beam, and (b) the formed plasma jets in which the high-intensity laser pump facilitates HHG. Reprinted with permission from Ref. 19. Copyright 2011. AIP Publishing LLC.

passing a high-intensity laser pulse through such plasma patterns suitable conditions for the QPM required for HHG can be created. These measurements suggest that such conditions exist between 140 ns and 180 ns after the initiation of plasma by the ablating laser pulse (Fig. 5.3). Within this temporal window the plasma jets are several hundred microns wide and have relatively uniform temperature and relatively low electron density of nearly 10^{17} cm^{-3}, whereas at later times the plasma structure begins to fade. Examining Fig. 5.3 in depth shows that the modulation of the plasma density is significant, although the authors suggest that much finer periodicities suitable for generation of higher harmonics could be obtained by using finer lithography in preparation of the target. They have demonstrated the feasibility of a robust scheme for tailoring plasma structures with control over material composition, temperature and density (both of free electrons and neutrals), through the ablation of specifically prepared lithographic targets, which can support the quasi-phase matched HHG. Notice that this approach yet found practical application in plasma harmonic generation.

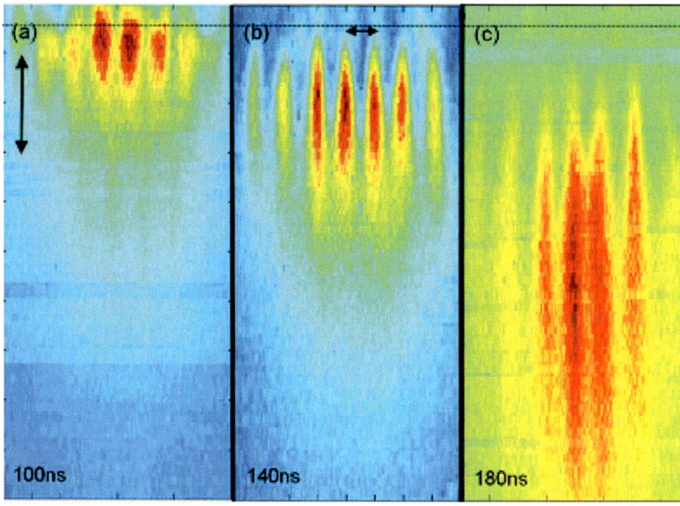

Figure 5.3 Monochromatic imaging of the plasma jets at different times using a 30 ns gate. The dashed line marks the target's surface, the double arrows are 200 mm scale. Reprinted with permission from Ref. 19. Copyright 2011. AIP Publishing LLC.

5.2 Stable Generation of High-Order Harmonics of Femtosecond Laser Radiation from Laser-Produced Plasma Plumes at 1 kHz Pulse Repetition Rate

Almost all HHG studies from weakly ionized plasmas produced during laser ablation of various solid targets were carried out using the 10 Hz pulse repetition rate lasers. Up to now, only few studies of HHG from plasmas have been carried out on static targets using 1 kHz class lasers [21–23]. The ablation process at 1 kHz pulse repetition rate causes a considerable change of the surface properties of the target due to the melting, which deteriorates the plasma plume formation during laser ablation. Surface heating and melting of a static target results in an unstable harmonic signal so that movement of the target surface is required to maintain a reasonable stability. The demand in finding the optimal way for improving the plasma harmonic stability at 1 kHz pulse repetition rate is high due to recent observations of the advanced properties of plasma harmonics over gas harmonics [9, 11, 24]. In particular, in [9], the plasma HHG conversion efficiency was measured one order of magnitude stronger than gas HHG efficiency. Analogous features were reported in [24].

The obstacle of all plasma harmonic experiments during earlier studies was an insufficient stability of plasma parameters (density, ion and free electron concentrations, excitation conditions, and so on), which led to an instability of harmonic yield and fast decay of harmonic efficiency during irradiation of the same spot of ablating target. As it was mentioned, most of those early studies were performed using a 10 Hz class lasers. At this relatively low pulse repetition rate, the stability of harmonics deteriorated after a few hundred shots on the same spot of the surface and even quicker for powder-like materials (fullerenes, nanotubes, metal nanoparticles, various organic and non-organic powders). One can note that laser ablation of the latter samples can be considered as an important tool for their structural studies using XUV nonlinear spectroscopy. The application of soft ablation allows the use of the same target for a much longer period than in the case of earlier studies of over-excited targets during laser ablation. Thus, a search of a robust, easy-to-apply method for improving the harmonic stability in the case of plasma HHG could considerably advance a search of the fundamental (structural, orientational, and so on) properties of organic and inorganic atoms and molecules.

The earlier used approaches of a rotating disc geometry [25–27] for the movement of targets during ablation are not suitable, as the distance between driving femtosecond beam and target surface should be maintained minimal (of order of 100 μm), while the Rayleigh length of the driving beam is maintained at the range of few millimeters. Below, I describe a new method using a motorized rotating rod specifically prepared for the HHG from plasma plumes using high pulse repetition rate lasers and demonstrate that this technique significantly improves the stability of high-order harmonics [28].

Those studies were performed using two laser pulses: one to produce the plasma plume and the second to drive the HHG within it. The first (heating) pulse was created by splitting off a portion (200 μJ) of the uncompressed 8 ps laser pulse from a 1 kHz

Ti:sapphire laser. The remaining pulse was compressed in a prism compressor and then further compressed using a hollow core fiber and chirped mirrors, resulting in 250 μJ, 3.5 fs pulses. The driving (probe) pulse was delayed with respect to the heating pulse by 40 ns to give the plasma the time to expand away from the target surface to allow the driving pulse to pass through the plasma without being clipped by the target.

Target rotation apparatus consisted of three linear stages driven by stepper motors along three axes. The target was attached to an axis of the fourth motor, which provided rotation with a variable speed [from a few rotations per minute (rpm) up to 300 rpm]. Rotating the target was sufficient to achieve stable harmonic radiation and an additional vertical movement was not required, though this capability might be useful for future plasma HHG experiments. As the setup requires the target to be positioned very close to the driving beam, it was of paramount importance that the target was carefully aligned to the axis of rotation. Any movement of the target surface due to eccentricity in the radial direction from the driving beam axis would result in an oscillation of the harmonic signal due to variation of the plasma density seen by the driving beam or, in the extreme case, clipping of the laser beam.

The target (metallic cylindrical rode with diameter of 10 mm and length of 30 mm) was positioned as shown in Fig. 5.4, with

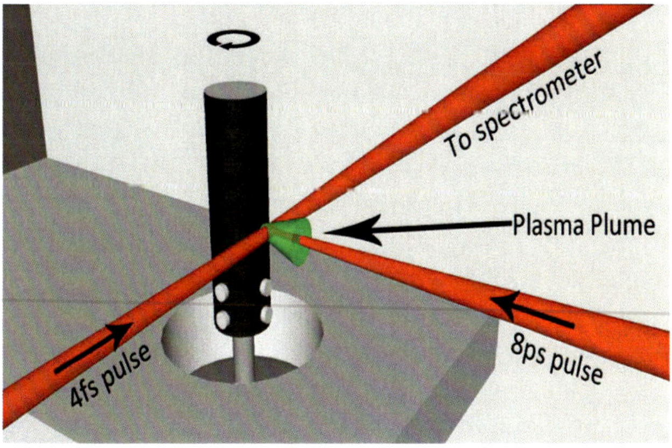

Figure 5.4 Schematic of the rotating target and HHG configuration. Reproduced from Ref. 28 with permission from Optical Society of America.

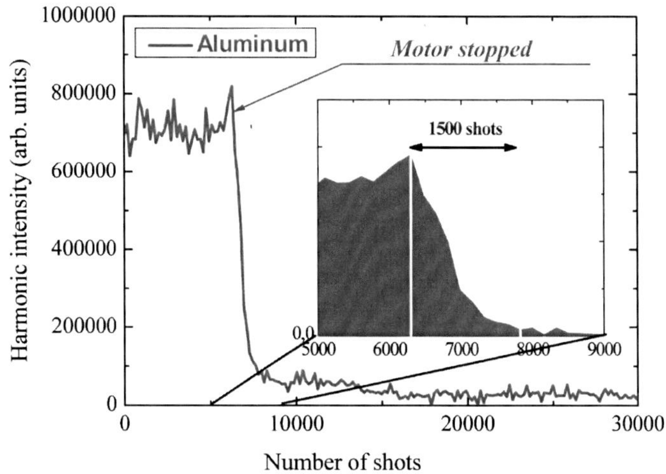

Figure 5.5 Decay of the harmonics from aluminum plasma after stopping the rotation of the motor. The harmonics were integrated over 40–80 nm spectral range. Reproduced from Ref. 28 with permission from Optical Society of America.

the driving pulse propagating 100–200 μm above the target surface. The picosecond heating pulse was focused onto the surface of the rotating target. In order to efficiently produce high-order harmonics, the plasma must be weakly ionized [6]. To achieve this, the target was positioned slightly in front of the focus of the heating pulse using a 50 cm focusing lens, leading to an on-target intensity of nearly 1×10^{10} W/cm^2. This also had the benefit of increasing the size of the plasma produced from ablating a larger area. The size of the focus at the target surface was measured to be ≈500 μm. The delayed driving pulse was focused through the plasma using a 40 cm spherical mirror. The HHG radiation was analyzed by an XUV spectrometer consisting of a flat-field grating and an imaging microchannel plate detector with phosphor screen imaged onto a CCD camera.

Figure 5.5 shows that there is a drastic change in the harmonic signal (integrated over the spectral range of 40–80 nm) when the rotation of the aluminum target is stopped. There is a sharp intensity decrease of more than one order of magnitude over only 1000 shots (or just after 1 s of ablation using 1 kHz laser). The benefits of the rotating rod are clearly shown in Fig. 5.6 in which stable harmonic generation was achieved from the plasma produced on an aluminum

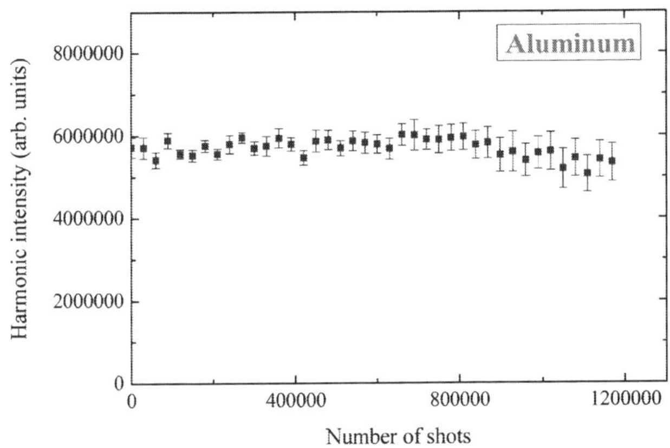

Figure 5.6 Stability of integrated harmonic signal from the aluminum plasma in the case of rotating target. Reproduced from Ref. 28 with permission from Optical Society of America.

target for over 1 million laser shots. Stable harmonics were achieved in a broad range of the speeds of rotation (from 10 rpm and faster). The target rotational speed and the size of the ablation focus imply that the same area of target was undoubtedly used repeatedly for consecutive rotations over the 20 min duration of experiments. This could result in thermal damage issues with this high pulse repetition rate.

It is possible that once the fixed surface is melted, the force from a following laser shot and plasma creation could expel some of the liquid target from the ablation area, which would not cause the plasma to be emitted in a direction normal to the surface. These effects are considerably diminished once the target starts to rotate. During rotation, the previously ablated area cools down such that, during the next set of ablation on this spot, the plasma formation occurs at approximately same conditions. To prove that the ablated area cools down with rotation, the target was rotated at different speeds (from 10 rpm to 300 rpm) and no significant difference in stability of harmonic yield was found. These observations point out the importance of the periodic change of the ablation zone. This also confirms a suggestion that the cooling of the ablation area leads to stable plasma generation.

Characteristics of plasma (density and ionization state) are the most important parameters to achieve and maintain stable

HHG efficiency during an extended period of illumination. The calculations [24] have shown that the concentration of particles in the area of femtosecond laser-plasma interaction at optimal delay between heating and probe pulses (~40 ns) is in the range of $2 \times 10^{18}\text{cm}^{-3}$. The solid surface was considered as the one unheated before the laser ablation. Indeed, after one round of rotation (e.g., after 0.2–2 s), the plasma disappears, the ablated spot cools down, and the next laser shot on the same spot can be considered as a shot on the almost "fresh" surface. Contrary, in the case of a stationary target, the following shots continue the heating of the same spot.

The novelty of this approach includes the observation of advanced properties of plasma HHG even at extremely small energies of the heating pulses (0.2 mJ). The efficiency of plasma HHG depends on the possibility to create "optimal" plasma. This can be done using both multi-mJ pulses, as was shown in previous studies [4], and few hundred μJ pulses, as it was demonstrated in the reviewed work [28] and recently published studies [24]. The important point here is the intensity and fluence of the heating pulse on the target surface. The application of a higher energy heating pulse could create the conditions of "optimal" plasma over a longer distance, which could (or could not) increase the harmonic yield depending on the phase relations between the driving and harmonic waves. In addition, it can also lead to the over-heating of the target at 1 kHz ablation. As it was already mentioned earlier, the rotating speed did not influence the stability of harmonics using 0.2 mJ heating pulses. The use of more energetic pulses at high repetition rate (i.e., of order of few mJ) may require additional optimization of the rotation target technique (e.g., by periodic up and down dragging of the rotating target).

5.3 High-Order Harmonic Generation in Graphite Plasma Plumes Using Ultrashort Laser Pulses: A Systematic Analysis of Harmonic Radiation and Plasma Conditions

The characteristics of laser plasma play a crucial role in determining how efficiently high harmonics can be generated in the plasma plumes. An increase in the free electron density was likely to have been the limiting factor for the harmonic cutoff energy in early experiments with laser plasmas. A search for appropriate target materials, which can provide favorable ablation plasmas for efficient HHG, has motivated the analysis of plasma characteristics at conditions of the high yield of harmonics. Recent studies have shown that carbon ablation plasmas are the promising media to satisfy the above requirements.

Shot-to-shot stability of the harmonic signal is crucial for any application of the generated radiation and especially for the measurement of the pulse duration of converted XUV radiation. Such temporal measurements were reported in the case of HHG in chromium plasma [9]. Apart from its fundamental interest, high-order harmonic generation in plasma plumes could thus provide an intense source of femtosecond and attosecond pulses for various applications.

Optical parametric amplifiers (OPAs) operating in the mid-infrared (MIR) range are promising tools for harmonic cutoff extension and attoscience experiments. The spectral cutoff energy of HHG obeys the scaling law $E_c \sim I\lambda^2$, where I is the peak intensity of the driving field and λ its central wavelength, which allows one to extend the harmonic emission beyond the 100 eV range by using longer wavelength laser sources. Another advantage of MIR OPAs is their wavelength tunability, which allows one to tune the spectral position of harmonics toward the ionic transitions with strong oscillator strengths. This feature allows the observation of resonance-enhanced harmonics and broadens the range of plasma samples where this phenomenon could be realized compared with the case of nearly 800 nm lasers of essentially fixed wavelength [7]. Moreover, by using two-color HHG techniques, the application of MIR OPAs allows the study of complex molecules during their ablation and HHG using the tunable long-wavelength radiation. These features are interesting for spectroscopic applications of HHG in the MIR range [29, 30].

In the meantime, the use of MIR OPAs for HHG should lead to a reduced harmonic generation efficiency that scales as λ^{-5} [31, 32]. It is of considerable interest to analyze the relative behavior of plasma harmonics in the cases of 800 nm and MIR lasers and thereby to find the conditions when the reduction of harmonic yield becomes not so dramatic due to some enhancement mechanisms, such as the presence of in situ produced nanoparticles, which increase the HHG conversion efficiency. It is worth noting that previous studies of plasma HHG in carbon plumes [11] have inferred, through analysis of plasma debris morphology, the formation of nanoparticles during laser ablation of carbon-contained targets.

Atomic carbon is a reactive species, which stabilizes in various multi-atomic structures with different molecular configurations (allotropes). All the allotropic forms of carbon (graphite, diamond, and amorphous carbon) are solids under normal conditions, but graphite has the highest thermodynamic stability. Laser ablation of graphite has been intensively examined during the last 10 years to define plasma conditions for the synthesis of carbon structures with unique properties. The physical characteristics of the plasma plume, such as concentration of atoms and clusters, directly affect the properties of the material being formed in the dynamic expansion of the ablated material. The successful synthesis of clusters is strongly dependent on the formation of atomic and molecular species with the required chemistry and aggregation ability. Thus, to select the optimal plasma conditions for HHG, a detailed understanding of the basic physical processes governing the ablation plume composition and reliable methods for controlling of the plume species are needed. The reasons mentioned above and the consideration of recent studies of HHG in carbon plasmas [11], as well as recently reported comparisons of the HHG in graphite-ablated plasmas and argon gas [24, 33], have prompted to systematically analyze the plasma conditions for optimal HHG conversion efficiency in graphite plasmas [34].

5.3.1 *HHG in Carbon Plasma at Different Conditions*

High-intensity few-cycle pulses (760 nm central wavelength, 0.2 mJ, 3.5 fs pulse with 1 kHz repetition rate) were typically obtained from the Ti:sapphire laser after second stage of compression consisting of hollow fiber filled with neon and bunch of chirped mirrors [35]. The compressed pulses were characterized with a spatially encoded arrangement for direct electric field reconstruction by spectral

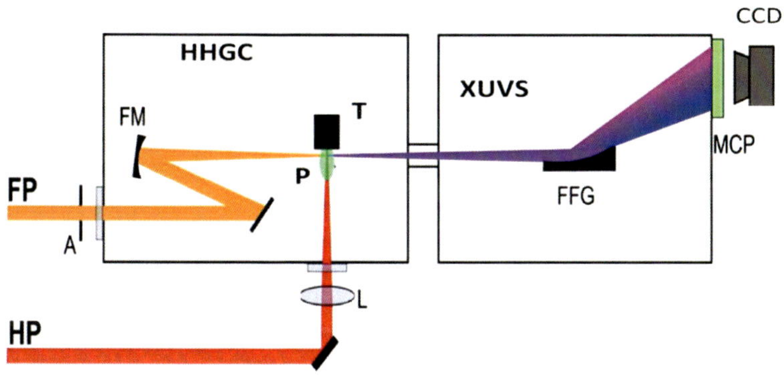

Figure 5.7 Experimental setup for harmonic generation in plasma plumes. FP: femtosecond probe pulse, HP: picosecond heating pulse, A: aperture, HHGC: high-order harmonic generation chamber, FM: focusing mirror, L: focusing lens, T: target, P: plasma, XUVS: extreme ultraviolet spectrometer, FFG: flat field grating, MCP: microchannel plate and phosphor screen detector, CCD: CCD camera. Reproduced from Ref. 34 with permission from IOP Publishing.

shearing interferometry. This radiation was used for frequency up-conversion in the specially prepared carbon plasma. A portion of the uncompressed radiation of this laser (central wavelength 780 nm, pulse energy 120 μJ, pulse duration 8 ps, pulse repetition rate 1 kHz) was split from the beam line prior to the laser compressor stage and was focused into the vacuum chamber to heat the graphite target and create a plasma on its surface (Fig. 5.7). These picosecond heating pulses were focused by a 400 mm focal length lens and created a plasma plume with a diameter of nearly 0.5 mm using an intensity on the target surface of $I_{ps} = 2 \times 10^{10}$ W/cm^2. The delay between plasma initiation and femtosecond pulse propagation was fixed at 33 ns. As an alternative ablation, the 10 ns, 1064 nm pulses from a 10 Hz repetition rate Q-switched Nd:YAG laser were used that provided an intensity on the target surface of 1×10^9 W/cm^2. In that case, the delay between the 10 ns heating pulses and the 3.5 fs probe pulses was varied in the range of 10–60 ns to maximize the harmonic yield.

The 3.5 fs probe pulses propagating in a direction orthogonal to that of the heating pulse were focused into the laser plasma using a

400 mm focal length reflective mirror. The position of the focus with respect to the plasma area was chosen to maximize the harmonic signal, and the intensity of femtosecond pulses at the plasma area at these conditions was estimated to be $I_{fs} = 6 \times 10^{14}$ W/cm^2. The 30 fs, 780 nm, 2 mJ probe pulses from another Ti:sapphire laser operating at 1 kHz repetition rate and producing approximately the same intensity inside the laser plasma were also used for HHG. The details of this setup and registration system are presented in [24, 36].

In order to analyze the harmonic yield of the MIR source in the graphite-ablated plasma an OPA pumped by the 30 fs Ti:sapphire laser was used. A beam splitter inserted before the laser compressor of this Ti:sapphire laser allowed to pick off 10% of the beam (780 nm, 1 mJ, 160 ps, 1 kHz pulses) to generate a plasma plume on the graphite targets, with the remaining 90% being compressed to 30 fs (7 mJ) to pump a commercial OPA. The OPA was optimized for high conversion efficiency, beam quality, and duration of the converted pulses. To achieve high reproducibility of the generated pulses, all the amplification stages were driven to saturation. This device generated 35 fs signal pulses in the 1200–1600 nm range. The idler pulses covered the 1600–2200 nm range. The delay between the heating ablation pulse and MIR pulses from the OPA was set to 35 ns, as this delay was found to be optimal for the efficient generation of extended harmonics.

As the goal of these studies was to analyze the graphite ablation plasma characteristics at the conditions of efficient HHG of ultrashort laser pulses, this process was firstly optimized by achieving the maximum conversion efficiency and highest harmonic cutoff using the driving radiation from both the Ti:sapphire lasers with fixed wavelengths and the tunable OPA. Then, the efforts were concentrated on the analysis of the "optimal" plasma plume using three techniques: optical emission spectroscopy of emitting plasma species in the visible, UV, and XUV spectral ranges; scanning electron microscopy for inspection of the deposited plasma debris; and finally time-of-flight mass spectrometry (TOFMS) for analysis of the ionic components of the plasma.

To analyze the influence of the spectro-temporal characteristics of the driving radiation on the harmonic yield in the case of ablation

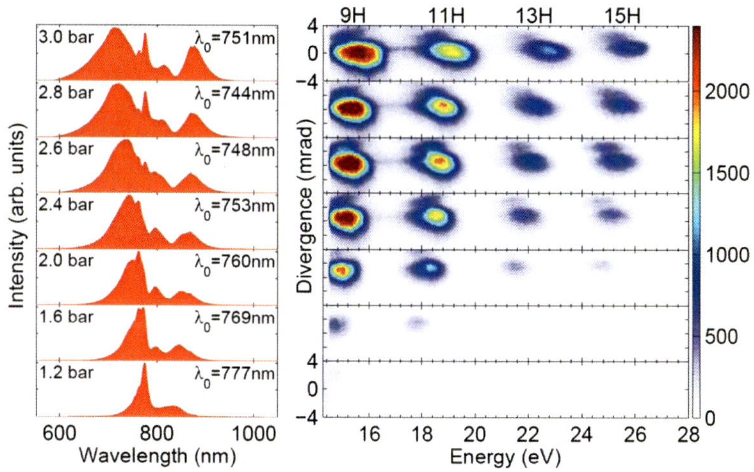

Figure 5.8 Carbon harmonic spectra as the functions of neon pressure in the hollow fiber. The corresponding laser spectra measured in front of the vacuum chamber are presented on the left side. The plasma was created using the 10 ns pulses. λ_0 is the central weighted wavelength of the spectral distribution. Reproduced from Ref. 34 with permission from IOP Publishing.

by 10 ns pulses, the backing pressure of neon in the hollow fiber of second compressor was changed, which allowed the variation of pulse duration from 25 fs to 3.5 fs [37]. The dependence of the spectral and intensity characteristics of the harmonic images recorded by the CCD camera in the 15–25 eV range at different input pulse spectra and backing pressures of neon are shown in Fig. 5.8. One can clearly see that, with the increase of backing pressure (from 1.2 to 3 bar), the harmonic intensity increases, while the harmonic wavelength spectrally shifts toward the blue. During these experiments, the driving pulse energy was held constant.

An interesting feature of the carbon harmonic spectrum from the 10 ns pulse-induced plasma is that the spectral width is about two to three times broader than that of harmonics generated in other atom-rich and ion-rich plasmas at the same fluence and intensity of heating pulse, when using few-cycle pulses. For example, the full width at half maximum for medium-order harmonics was 1.5 nm in the case of graphite plasma, versus 0.4 nm for different metal

(Ag, Al, and Cu) plasmas. The broader width of the harmonics can be explained by self-phase modulation and chirping of the fundamental radiation propagating through the carbon plasma. The presence of nanoparticles in the plasma plume may also contribute to bandwidth broadening of harmonics.

For practical applications of the coherent short-wave radiation generated in graphite plasma using a 1 kHz driving laser, it is necessary to analyze the stability of the plasma characteristics and the generated harmonics. Recently introduced new technique for maintaining a stable ablation plasma for harmonic generation using high pulse repetition rate lasers (>1 kHz) based on a cylindrical rotating target [28] was described in previous section. The studies under discussion have shown that, in spite of different properties of metal and graphite targets, the rotating target allowed achieving stable HHG in both metal and graphite plasmas. The rotating graphite rod allowed maintaining a relatively stable harmonic yield well above 1×10^6 laser shots. Harmonics up to the 29th order were routinely observed in those studies using the 3.5 fs pulses.

Upper panel of Fig. 5.9 shows the harmonic spectrum generated in the case of 1300 nm probe pulses. Harmonics up to the 57th order were observed at the conditions of carbon plasma formation using the heating uncompressed 160 ps pulses from this laser. It is worth noting that application of less intense 1400 nm pulses available by tuning the OPA, while generating weaker harmonics, did not result in a higher harmonic cutoff than in the case of 1300 nm. This observation suggests that the harmonic generation occurred under saturated conditions, with the expectation of even stronger harmonics once the micro and macro-processes governing frequency conversion are optimized.

Harmonic spectra up to the 29th order in the case of 780 nm, 30 fs driving pulses are presented in the bottom panel of Fig. 5.9. By comparing with the spectra collected with the 1300 nm driving source (same figure, upper panel), one can clearly see the expected extension of harmonic cut-off in the case of the longer wavelength driving source. The important peculiarities of these comparative studies are the broadband harmonic spectra in the case of 1300 nm laser and the similar yield of harmonics at the two driving wavelengths. Although the former feature depends on

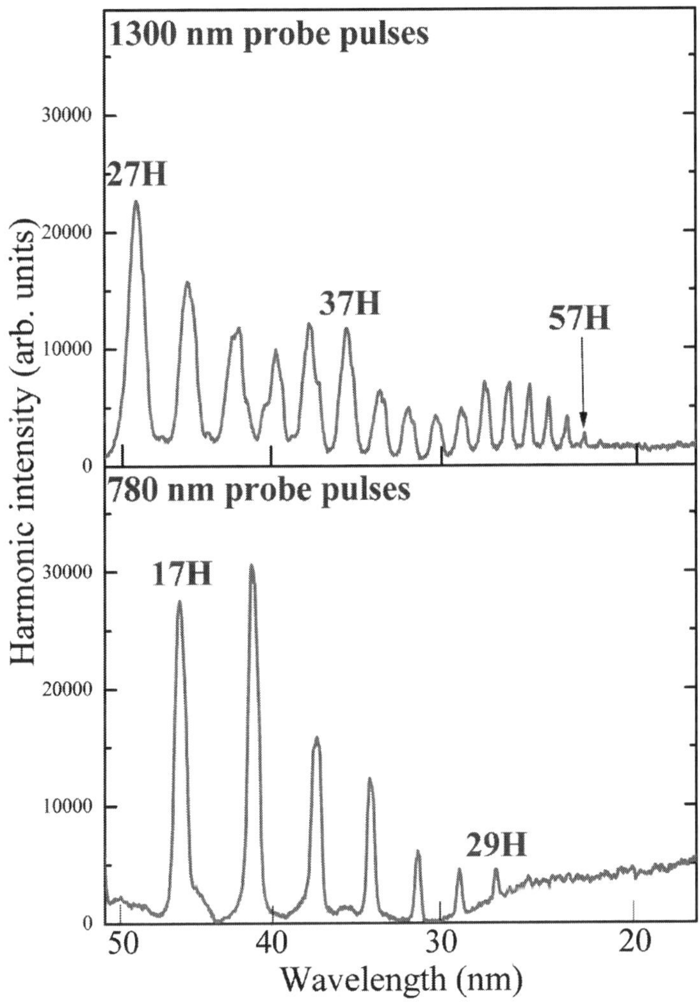

Figure 5.9 Carbon plasma harmonic spectra using the 1300 nm (upper panel) and 780 nm (bottom panel) probe pulses. The energies of probe pulses were 0.2 mJ (upper panel) and 0.54 mJ (bottom panel). Ablation was carried out using 160 ps, 780 nm, 1 kHz laser pulses. Reproduced from Ref. 34 with permission from IOP Publishing. All rights reserved.

the bandwidth of the OPA output, the latter observation requires additional consideration. The plasma harmonic yield from the MIR source did not follow the expected $I_{\rm h} \propto \lambda^{-5}$ rule. In fact, for the intensities of MIR and 780 nm pulses used [$\sim$ (2-4) $\times$ 10^{14} W/cm^2], the harmonic efficiency of the XUV radiation driven by MIR pulses was higher than the case of 780 nm pulses, while using lower energy of the former pulses (0.2 and 0.54 mJ respectively). One can note that the $I_{\rm h} \propto \lambda^{-5}$ rule predicts a nearly 13-fold decrease of conversion efficiency for the MIR (1300 nm) pulses compared with the 780 nm pulses at equal probe pulse intensities.

5.3.2 *Characterization of Optimal Plasma Conditions*

This subsection presents the characterization of the graphite ablation plasma plumes at conditions of maximum HHG conversion efficiency. In graphite, the ablation plasma plume may contain various species of carbon, that is, neutrals and ions, small molecules, clusters, aggregates, and so on, which can contribute to harmonic generation in various extents. It is important to determine their presence in the region where the driving laser pulse interacts with the expanding plasma. In particular, the production of clusters in the laser plasma during laser ablation of various targets has a high probability, although their presence and concentration in the plasma area where the frequency conversion occurs is yet to be confirmed directly. Another issue is how one can define the density of monomers, dimers, and clusters and their influence on the HHG yield. Analysis of post-ablation conditions of the deposited debris can provide information on the nature of those species, despite the differences between the composition of the plasma in its early stages and the deposited material, due to the influence of aggregation on the substrate [38]. Another issue of interest is whether the spectral characterization of the plasma emission in the visible and UV ranges can provide some clues about the plasma conditions, without the simultaneous analysis of the XUV emission.

Plasma characterization through optical spectroscopic measurements in the visible, UV, and XUV ranges at the conditions of different HHG efficiencies was carried out using the above-described XUV spectrometer and a fiber spectrometer. The fluences of heating pulses at which the spectra were recorded corresponded to both optimal and non-optimal conditions of HHG. The characterization of the plasma debris collected on the silicon wafers placed 4 cm from the ablated target was carried by scanning electron microscopy.

Cluster composition of the ablation plume produced by nanosecond laser pulses was investigated by TOFMS. A brief description of the experimental set up is given here; more details of the TOFMS can be found elsewhere [39]. The laser beam (1064 nm, 5 mJ, 10 ns pulse duration) was focused at normal incidence to a 0.2 mm spot on the surface of the graphite target. The laser intensity on the target surface was 1.5×10^9 W/cm^2, which resulted in the creation of an optimal plasma for efficient HHG using the nanosecond ablation pulses. The target was placed in a vacuum chamber (pumped to $\sim 2 \times 10^{-6}$ bar) between the extracting and accelerating plates of a linear TOFMS. The target surface was parallel to the flight axis of the spectrometer. The target could be rotated and displaced at variable distances from the axis. Positive ions produced in the ablation were deflected along the TOFMS axis by an electric field typically in the range of 300–400 V/cm and accelerated by a total voltage of 2500 V. A high-voltage switch was used to apply the bias voltage at controlled delays with respect to the laser ablation pulse. Ions entered the drift region (flight length ~1 m) and were detected by a microchannel plate. Analysis of neutral species produced in the ablation could also be performed by the use of a second post-ionization laser source (F_2 excimer laser irradiated at $\lambda = 157$ nm). The post-ionization laser pulse interacted with the ablation plume perpendicularly to the plume propagation axis, at different distances from the target surface and at different delays with respect to the ablation laser.

It has been shown previously that efficient harmonic emission is observed only in the case when the visible and UV plasma emission is dominated by neutral and singly ionized carbon lines [6]. The present studies have also confirmed this feature at the laser fluence used to heat the target surface. The broad features of emission spectra near 470, 515, and 555 nm could be assigned to the bands of excited C_2 molecules. These bands have also been observed in early studies of the ablation of graphite (see for instance [40–42]). Other lines in the spectra were attributed to the neutral and singly charged carbon.

The analysis of optical spectra in the visible and UV ranges does not provide information about the presence of highly ionized species, which can be revealed by collecting the plasma emission in the XUV range. The XUV spectrum of carbon plasmas collected following excitation by a 8 ps heating pulse at high intensity, without further excitation by the driving pulse, provides some insight into the plasma components prior to interaction with the driving radiation. This spectrum was collected under conditions of considerable decrease of the nonlinear optical response of the medium (i.e., at weak HHG) and revealed the appearance of many emission lines from C II and C III ions. Over-excitation of the target by 10 ns pulses also led to the appearance of emissions from high-charged (C III, C IV) ions.

It may be noted that those measurements were time-integrated, so one could not say exactly which plasma components existed at the moment of the propagation of the femtosecond beam through the plume. However, the presence of ionic lines from multi-charged species in the last two cases gives a strong indication of over-excitation of the target and of its negative influence on HHG efficiency. One can note that, at this level of excitation of the graphite plasma, harmonic generation was partially or entirely suppressed. Specifically, a two-fold increase in the intensity of 8 ps pulses (from 2×10^{10} to 4×10^{10} W/cm^2) led to a decrease of harmonic intensity by a factor of 2.5. The same can be said about the excitation using longer (10 ns) pulses, though the threshold, at which harmonics started to decay, was considerably lower (2×10^9 W/cm^2). Application of 10 ns pulses with an intensity of 3×10^9 W/cm^2 led to a substantial decrease of harmonic efficiency and to the appearance of emission lines from high-charged ionic species. One can note that, under conditions of efficient HHG, no ion lines appear alongside the harmonic spectra.

To prove the presence of clusters in carbon plasmas under optimal conditions of harmonic generation, the morphology of deposited debris from the graphite-ablated plasma was analyzed during ablation by picosecond and nanosecond pulses. It has already been mentioned that laser ablation of a solid material is a widely accepted technique for the generation of nanoparticles. However, this process has previously been studied without taking into account the role of free electrons and highly excited ions, which destroy the optimal conditions for phase-matched HHG. SEM measurements of the deposited debris were carried out at the laser ablation conditions corresponding to the optimal plasma formation for efficient HHG. The substrates (glass plates and silicon wafers) used to collect the deposited material were placed at a distance of 40 mm in front of the ablation area and the debris was further analyzed by SEM.

At optimal plasma conditions created by 8 ps pulses, when the highest harmonic conversion efficiency from the carbon-containing plasma was measured, the SEM images did not reveal the presence of nanoparticles in the deposited debris with sizes above the limit of detection (5 nm) of the microscope. This was probably due to the small fluence (0.2 J/cm^2) of the heating radiation on the target surface ($I_{ps} = 2.5 \times 10^{10}$ W/cm^2). It is possible that in the case of the carbon plasmas produced at these conditions, harmonics could also originate from nanoparticles with sizes below the limit of detection. Another pattern was observed upon ablation of the target with 10 ns pulses, where considerably higher heating fluence (10 J/cm^2) caused the appearance of nanoparticles on a nearby substrate. At relatively moderate conditions of ablation using 10 ns pulses [$I_{ns} \approx (1\text{–}2) \times 10^9$ W/cm^2], a high density of nanoparticles appeared in the SEM images of the deposits, with sizes mostly distributed in the range between 10 and 200 nm, with a mean size of 50 nm (Fig. 5.10a). One has to reiterate that debris characteristics were measured at maximum conversion efficiency of 15th-23rd harmonics. These morphological studies have confirmed the presence of relatively large nanoparticles deposited on the substrates at conditions of "optimal" ablation using 10 ns pulses. However, some uncertainty still remains about the correlation of these results with the presence of the same nanoparticles in the carbon plasma during harmonic generation, due to the possibility of aggregation of these clusters after deposition. To address this issue, the TOFMS of nanosecond pulse-ablated graphite was carried out.

Figure 5.10b shows the mass-resolved spectrum of carbon plasma after 60 shots of the 10 ns heating pulses. Those studies have revealed that, at plasma conditions close to optimal for HHG, the laser plume contains a group of small, singly ionized carbon clusters

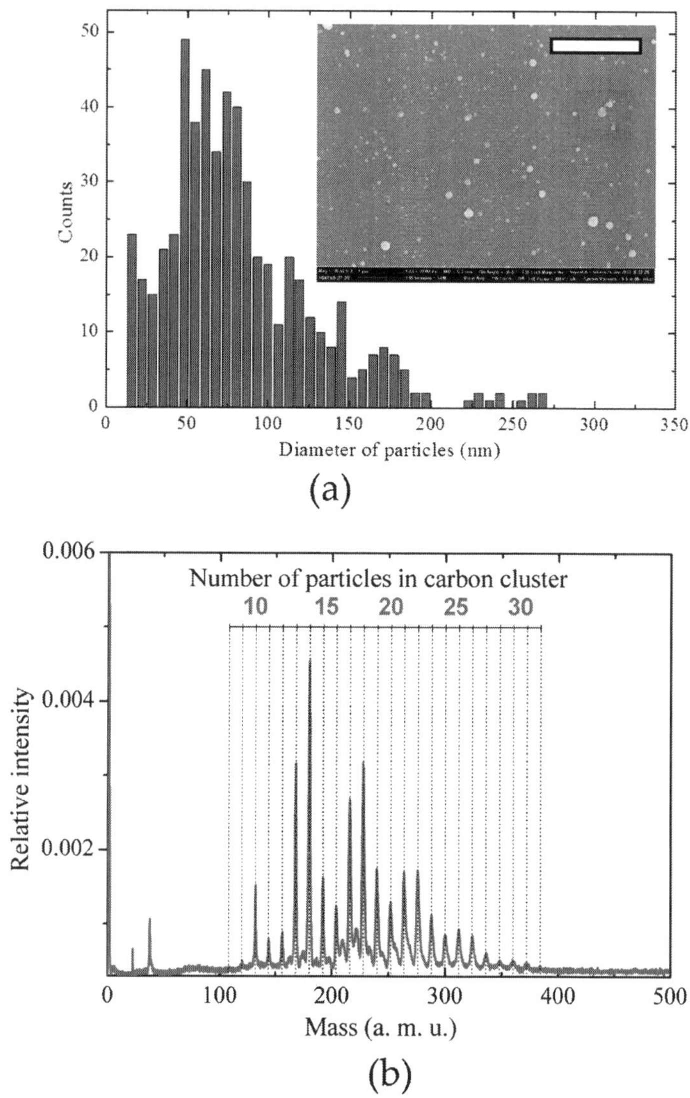

Figure 5.10 (a) Histogram of size distribution of deposited nanoparticles and corresponding SEM image of deposited debris in the cases of ablation of a graphite target using 10 ns pulses. The size bar on the SEM image is 2 μm. (b) Mass spectrum (in atomic mass units) of carbon plasma obtained during excitation of a graphite target using 10 ns pulses at the conditions close to optimal plasma formation for HHG. Reproduced from Ref. 34 with permission from IOP Publishing. All rights reserved.

(C_{10}–C_{30}). The attempts to find higher mass clusters failed, though a search for them was carried out over a longer range of delays (up to few μs) between the onset of laser ablation and the switching on the triggering pulse in the TOFMS.

To ascertain the presence of neutral species in the ablated plasma using TOFMS, an additional source of ionization should be used. In the discussed case, an F_2 post-ionization excimer laser (λ = 157 nm) served that purpose. Photons of this wavelength can induce the ionization of neutral clusters through single-photon absorption. However, once the 157 nm pulse was focused on the plasma area, no evidence in the mass spectra of neutral clusters was found. This could be explained by the fact that the ionization potential of carbon (I_p = 11.2 eV) is higher than the photon energy of the ionizing laser (E = 7.9 eV) and the difficulty of ionizing the existing carbon clusters by two-photon absorption of 157 nm radiation, due to low intensity of these pulses in the plasma area.

The concentrations of carbon plasma at the experimental conditions of target ablation (i.e., 2×10^{10} W/cm^2 in the case of 8 ps pulses and 1×10^9 W/cm^2 in the case of 10 ns pulses) allowing efficient harmonic generation were calculated using a three-dimensional molecular dynamical simulation of laser ablation of graphite using the molecular dynamics code ITAP IMD [43]. The corresponding concentrations were found to be 2.6×10^{17} cm^{-3} and 2.5×10^{18} cm^{-3}.

5.3.3 *Analysis of HHG and Plasma Characteristics*

An earlier study has suggested that the presence of nanoparticles in carbon laser ablation plasmas can explain the observed strong harmonic yield from these media [11]. It was reported that the debris from ablated graphite and carbon lead targets contained nanoparticles with sizes between 100 and 300 nm. The authors of [11] therefore suspected that nanoparticles formed in the plasma by ablation were the source of intense harmonics. Heterogeneous decomposition, liquid phase ejection and fragmentation, homogeneous nucleation and decomposition, and photomechanical ejection are among the processes that can lead to the aggregation and disintegration of nanoparticles [44–46]. A number of different techniques were used in those studies to determine the aggregation state of the evaporated material, including time-resolved emission spectroscopy, CCD camera imaging of the plasma plume, Rayleigh scattering, and laser-induced fluorescence.

In discussed studies [34], SEM for debris analysis and TOFMS were applied for plasma characterization. These two methods have provided useful clues about the conditions and dynamics of plasma plume formed above the target surface. Although the former method can provide information about the presence of nanoparticles in the plasma, one has to cautiously consider those results from the following point of view. The deposition process on the substrate happens much later than the time of HHG emission, and the physical process of deposition may lead to further aggregation. As SEM is an ex situ method, one cannot exclude the difference between the real composition of nanoparticles in the plasma and the results of SEM measurements, although it clearly proves the presence of clusters in the plasma. TOFMS yields information on the in situ presence of ionized clusters, although it requires ablation of the target at the same conditions as in the case of HHG experiments and is not suited for the detection of neutral nanoparticles in the ablated plasma.

TOFMS measurements did not reveal the presence of neutral clusters in the 10 ns pulse produced plasma for the reasons described in the previous subsection. However, other studies (see, e.g., [41]) have indicated the presence of neutral carbon clusters using two-photon ionization with an ArF laser (photon energy 6.4 eV). Early TOFMS studies of laser ablation of graphite have revealed the typical characteristics of the expanding plasma species (average velocity 1.5×10^5 cm/s) and their concentration (4×10^{18}–6×10^{19} cm^{-3} [47]) for ablation with 532 nm, 10 ns pulses at fluences of the order of 3 J/cm^2. The measured mass distribution shown in Fig. 5.10b, revealing the presence of C_{10} to C_{30} species, is in good agreement with those observed in previous studies of graphite-ablated plasma at similar excitation conditions [47]. The restriction of cluster sizes to small-sized carbon nanoparticles has also been reported in [48], where it was argued that stronger excitation conditions are necessary to observe clusters larger than C_{32}. In that case, one should expect the appearance of closed cages made of joined five and six member rings. It was confirmed that C_{60} and C_{70}

fullerenes are the most abundant species among the high-mass ions of the carbon plasma plumes at higher ablation fluences. It was also suggested [47] that it is very likely that the plasma is sufficiently dense for cluster growth to occur via ion-molecular reactions. The kinetic mechanism can be responsible for the formation of carbon cluster ions, as the supersonic entrainment method is expected to considerably cool down the cluster ions. The growth of clusters is based on the addition of many small carbon neutral species to the ions in a stepwise fashion.

An explanation for strong harmonic generation from nanoparticles compared with single atoms or ions could be the higher concentration of neutral atoms inevitably accompanying the presence of nanoparticles. The increase of electron recombination cross-section for clusters with respect to atoms can also potentially enhance the HHG efficiency in nanoparticle-contained plasmas. Earlier studies of HHG from gases [14, 30, 49], as well as from the plasmas containing various nanoparticles (Ag, Au, $BaTiO_3$, and so on) [6, 8], have proven these assumptions by demonstrating the enhanced HHG from clusters as compared with single atoms and ions. Further evidence of the cluster contribution to the enhancement of the harmonic generation comes from investigations of very intense laser ablation of a silver target [50], which gave the assumptions regarding the participation of in situ generated nanoparticles.

The observation of a strong extended harmonic plateau in the case of the 1300 nm driving radiation also suggests the involvement of clusters in the HHG process with MIR pulses. Assuming the expected decrease of harmonic intensity from single particle emitters with the growth of driving radiation wavelength ($I_h \propto \lambda^{-5}$, [31, 32, 51, 52]), one can anticipate at least one order of magnitude decrease of harmonic yield from MIR pulses as compared with the harmonic yield obtained with 780 nm radiation at other equal conditions, in particular, pulse energy and duration. However, the experiment did not show a considerable difference between the intensities of harmonics originated from these two driving sources (Fig. 5.9). The energy of the 1300 nm pulses in the plasma area (0.2 mJ) was lower than the Ti:sapphire pulse (0.54 mJ). This suggests the involvement of a mechanism, which compensates for the expected considerable decrease of harmonic efficiency for the longer wavelength laser. The involvement of a clustered component of laser plasma in the process of frequency up-conversion may arguably explain the observed inconsistence with the theoretical predictions of the $I_h \propto \lambda^{-5}$ rule defined for atomic species [53, 54].

In principle, the enhancement of the harmonic spectrum from the carbon plume in the 15–26 eV range invokes the involvement of surface plasmon resonances of nanoparticles, analogously to the case of the fullerenes [3] in the range of their giant resonance in the vicinity of 20 eV. To prove this in the case of carbon plasma, one should provide evidence of giant absorption in the above range, but this has not been reported yet in the literature. The plasmonic properties of carbon nanoparticles can be responsible for the observed enhancement of carbon harmonics; however, their role requires additional study. Another option for explaining the high harmonic generation yield from the carbon plume is the indirect involvement of the clusters in HHG that, although not participating as harmonic emitters, could rather enhance the local field, analogously to recently reported studies using the gold nanostructures enhancing gas HHG [53, 54].

As it was mentioned, recent comparative studies of lower order harmonic efficiency in argon gas and carbon plasmas have revealed stronger conversion efficiency in the carbon plasmas [24, 33]. In this subsection, we have discussed evidence of the superior properties of graphite ablation for HHG. Some arguments, which could explain the enhanced high-order harmonic yield from this medium, are as follows: (a) the graphite target allows easier generation of a relatively dense carbon plasma and the production of adequate phase-matching conditions for lower-order harmonic generation; (b) the first ionization potential of carbon is high enough to prevent the appearance of high concentration of free electrons, a condition that is not necessarily met in metal plasma plumes; (c) neutral carbon atoms dominate in the carbon plume at optimal conditions of HHG before the interaction with the femtosecond laser pulse; and (d) carbon species allow the formation of multi-particle clusters during laser ablation, which can enhance the HHG yield.

5.4 Isolated Sub-Femtosecond XUV Pulse Generation in Mn Plasma Ablation

In this section, I discuss the HHG from transition metal plasmas. These are very promising targets in view of the giant resonances found in the photoionization cross-sections. For example, the Mn^+ cross section is nearly 40 Mb at 50 eV photon energy [55], whereas rare gas atoms have cross-sections between 1 and 8 Mb at this photon energy [56]. Photorecombination–the third step in the recollision model–is the inverse process of photoionization [57] and therefore HHG and photoionization must exhibit the same resonances. This has been confirmed not only by previous resonance-induced experiments with laser-produced transition metal plasmas but also in a recent study of HHG from xenon gas [58].

Resonance-induced enhancement of a single harmonic of the laser radiation allowed considerable improvement of harmonic efficiency in some specific XUV spectral ranges related with high oscillator strengths of ionic states of metals. This was confirmed in multiple studies following the initial observation of this phenomenon in indium plasma [59]. In particular, strong enhancement of a single harmonic was reported in Cr [60] and Mn [61] plasmas. The Mn plasma is of special interest, as it showed the highest harmonic cutoff energy observed in plasma plumes (101st harmonic of Ti:sapphire laser [61]). In previous studies, multi-cycle (30 [61] and 140 fs [62]) laser pulses were employed and the generation of all harmonics in the plateau was observed together with a strongly enhanced harmonics in the vicinity of 50 eV.

Recent progress in the generation of few-cycle pulses allowed the observation of various new effects including the realization of isolated attosecond pulse generation in gas media [63–65]. In this connection, it is interesting to analyze resonance-induced processes observed in an ablation plume using the shortest available laser pulses. Below, I present the analysis of the experiments on resonance enhancement in manganese plasmas using 3.5 fs pulses [66]. The most interesting feature observed in those experiments was a suppression of almost all neighboring harmonics in the vicinity of a resonantly enhanced single harmonic at the photon energy of nearly 50 eV.

The experimental arrangements were analogous to those presented in previous sections of this chapter. The harmonic spectrum in the case of propagation of 3.5 fs pulses through the manganese plasma was strikingly different compared with other plasma samples (e.g., Ag plasma) analyzed in separate experiments. Although all other samples studied showed a relatively featureless harmonic spectra with extended cutoff (Fig. 5.11, upper curve showing spectrum of the harmonics generating in the silver plasma), the Mn plasma allowed the generation of a strong single harmonic substantially enhanced compared with neighboring ones (Fig. 5.11, bottom curve).

As it was already mentioned, the harmonic spectra from manganese plasmas for 30 fs and 140 fs pulses also showed enhanced harmonics around 50 eV. The assumption of the resonance nature of the enhancement of harmonics of the nearly 800 nm radiation of Ti:sapphire lasers in this spectral region is supported by the presence of a strong giant resonance in the vicinity of 50 eV confirmed by experimental [55, 67] and theoretical [68] studies. The enhancement of a single harmonic can be attributed to the broadband resonances of the ions of few metals, such as V, In, Cd, Cr, Cd, and Mn. These "giant" resonances have been experimentally confirmed in the literature [55, 67, 69] and discussed recently in a few theoretical studies [70–73].

However, in previous studies using multi-cycle probe pulses, the intensity of enhanced harmonics was only a few times higher than those neighboring harmonic orders. The same features were reproduced in the reviewed studies using the 40 fs pulses from another Ti:sapphire laser at similar intensity inside the laser plasma (4×10^{14} W/cm^2). The raw image of the harmonic spectrum presented in Fig. 5.12a shows several enhanced harmonics starting from the 31st order followed by an extended second plateau. The extension of the harmonic cutoff exceeding the 71st order is attributed to the involvement of doubly charged Mn ions as the sources of HHG. This feature of Mn plasma harmonics has already been reported earlier [61]. A typical image of a Mn harmonic spectrum in the case of 3.5 fs pulses is also presented here (Fig. 5.12b). No second plateau, which was seen in the case of multi-cycle (40 fs) pulses, is observed for the few-cycle pulse. Most striking was

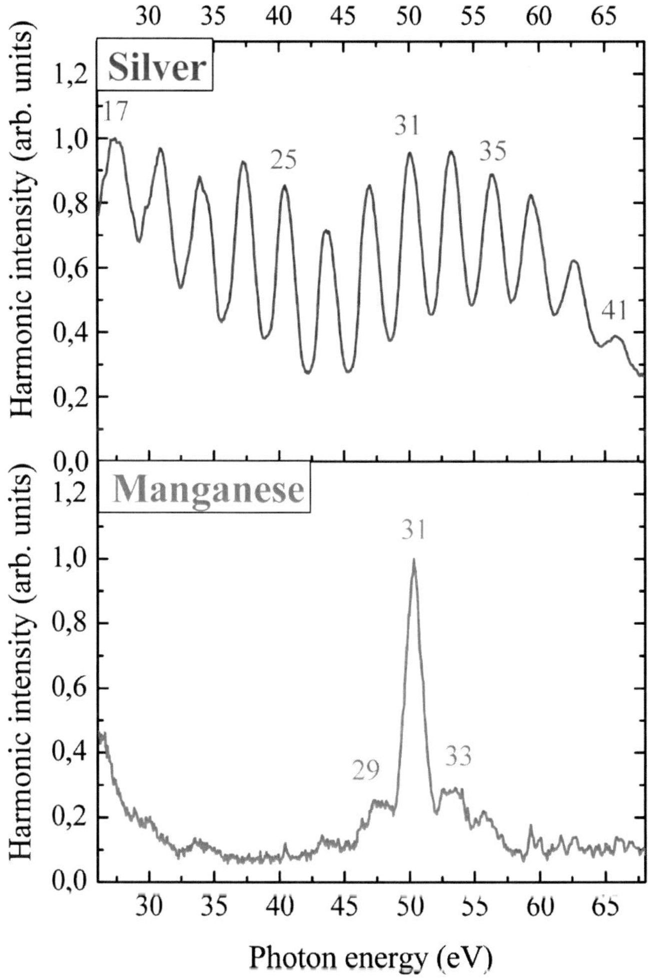

Figure 5.11 Harmonic spectra from the silver plasma (upper curve) and manganese plasma (bottom curve). Reproduced from Ref. 66 with permission from Optical Society of America.

the observation of a single, very strong, broadband (2.5 eV) 31st harmonic. Only two weak neighboring harmonics (around the strong emission) are seen in the 30-65 eV spectral range. The ratio between the intensities of the enhanced harmonic to the weak neighboring harmonics exceeded one order of magnitude. One can note that, at a

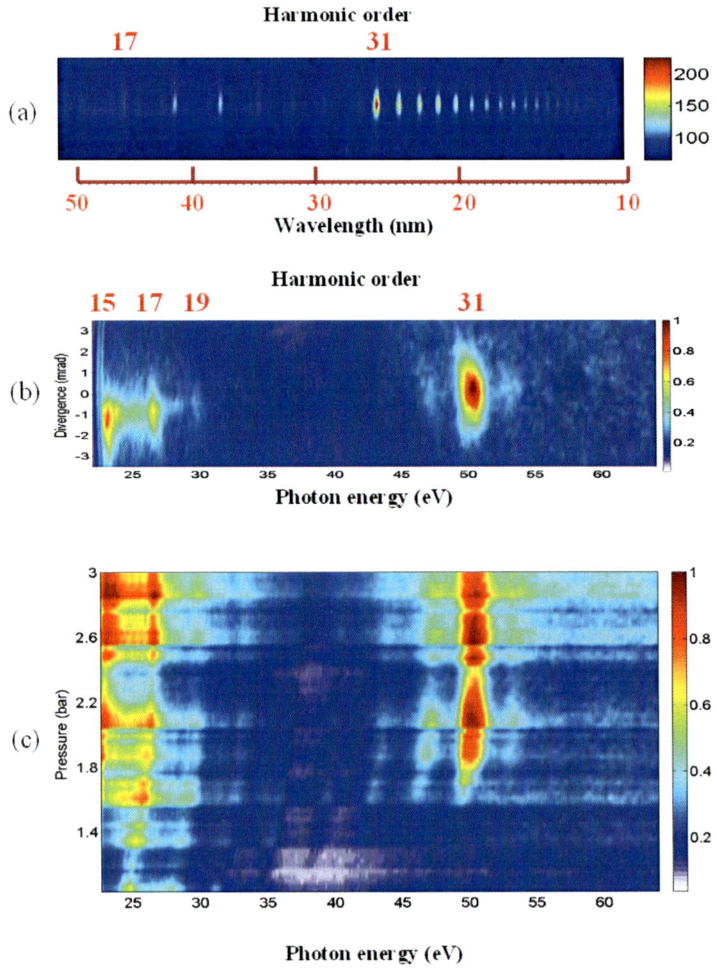

Figure 5.12 Raw images of the harmonic spectra from manganese plasma in the case of (a) 40 fs and (b) 3.5 fs driving pulses obtained at the same intensity. (c) Raw images of harmonic spectra from Mn plasma at different pressures of neon in the hollow fiber obtained at the same energy of driving laser pulses. Reproduced from Ref. 66 with permission from Optical Society of America.

lower intensity of the femtosecond pulse ($<2 \times 10^{14}$ W/cm^2), this strong harmonic disappeared when using both multi-cycle and few-cycle pulses.

The distinctive structure of the harmonic spectra, both for 40 fs and 3.5 fs pulses, clearly points out the involvement of Mn resonances centered around 50-51 eV. The same can be said about the photoionization or photoabsorption characteristics of Mn^+ plasma, which are due to the 'giant' $3p \rightarrow 3d$ resonance [55]. The laser polarization dependence of this emission was analyzed and it was found that the 50 eV radiation abruptly disappears with the change of the polarization state of the femtosecond driving pulses from linear to elliptical, which is a clear signature of the emission being due to high-order harmonic generation.

To analyze the effect of the spectro-temporal characteristics of the femtosecond radiation on the harmonic yield, the pressure of neon in the hollow fiber of second compressor was varied, thus changing the duration of the harmonic drive pulse [74]. The spectral and intensity variations of manganese harmonic spectra in the range of 22–62 eV as the functions of neon pressure in the hollow fiber are shown in Fig. 22c. One can clearly see that, with change of pressure (from 1 to 2.3 bar), the single 31st harmonic intensity varies from almost zero to its maximum high value. A blue shift of the harmonics is also evident. Further increase of neon pressure up to 3 bar, at which the experiments with the 3.5 fs pulses were carried out, did not change the harmonic distribution.

The experiments described above were carried out without carrier envelope phase stabilization (i.e., for random CEP values). The HHG experiments with Mn plasma using 3.5 fs pulses were also performed with stabilized CEP (e.g., at phase values of $\phi =$ 0 and $\pi/2$) and no considerable differences were found in that case (Fig. 23), though some variation of harmonic distribution was observed for the lower order harmonics (compare the middle and bottom curves of Fig. 5.13; see also Section 2.5). The spectral shapes of the 31st harmonic emission were quite similar for these two fixed values of CEP, while a considerable difference in harmonic spectra was maintained when comparing with longer pulse duration and lower intensity of the driving pulses. Figure 5.23 shows HHG

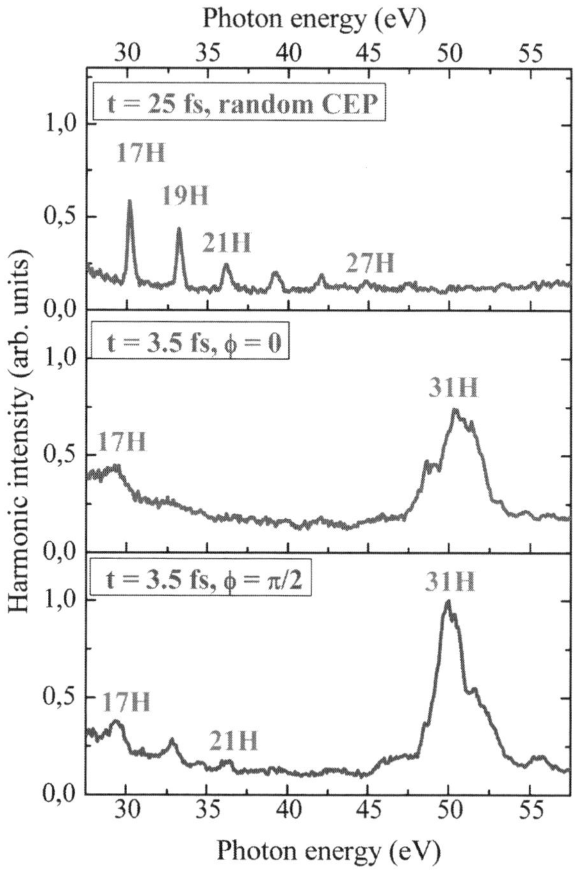

Figure 5.13 Experimental harmonic spectra generated from manganese plasma in the case of the absence of gas in the hollow fiber compressor ($t = 25$ fs) and random CEP (upper panel), and at 3 bar pressure ($t = 3.5$ fs) at fixed CEP ($\phi = 0$, middle panel; $\phi = \pi/2$, bottom panel). ϕ denotes the carrier envelope phase. Reproduced from Ref. 66 with permission from Optical Society of America.

measurements for 25 fs pulses (upper panel) and 3.5 fs pulses (middle and bottom panels) of the same energy. One can clearly see the absence of harmonic extension and resonance-induced HHG in the case of low-intensity, 25 fs pulses.

The fact that a strong CEP dependence of the plasma harmonic spectra in the case of 3.5 fs pulses was not observed could also be attributed to the presence of a significant amount of free electrons in the manganese plasma, which might diminish the difference between the HHG spectra recorded for different values of CEP. The same can be said about other HHG experiments using silver and brass plasmas, which did not show significant differences in harmonic spectra when comparing few-cycle pulses with fixed and random CEP. In the meantime, comparative studies with gas media under similar experimental conditions were carried out and found a characteristic dependence of the HHG spectra on the CEP commonly observed at such conditions. Thus, the absence of the considerable influence of the CEP on the harmonic pattern generated by few-cycle pulses from the ablation plumes appears to be a common feature of plasma HHG.

References

1. R. A. Ganeev, *Laser Phys.* 18, 1009 (2008).
2. R. A. Ganeev, *Open Spectrosc. J.* 3, 1 (2009).
3. R. A. Ganeev, *Laser Phys.* 21, 25 (2011).
4. R. A. Ganeev, *J. Phys. B: At. Mol. Opt. Phys.* 40, R213 (2007).
5. R. A. Ganeev, *Phys. Uspekhi* 52, 55 (2009).
6. R. A. Ganeev, *J. Modern Opt.* 59, 409 (2012).
7. R. A. Ganeev, *Laser Phys. Lett.* 9, 175 (2012).
8. R. A. Ganeev, *Laser Phys.* 22, 1177 (2012).
9. L. B. Elouga Bom, S. Haessler, O. Gobert, M. Perdrix, F. Lepetit, J. F. Hergott, B. Carré, T. Ozaki, and P. Salières, *Opt. Express* 19, 3677 (2011).
10. R. A. Ganeev, L. B. Elouga Bom, J. Abdul-Hadi, M. C. H. Wong, J. P. Brichta, V. R. Bhardwaj, T. Ozaki, *Phys. Rev. Lett.* 102, 013903 (2009).
11. Y. Pertot, L. B. Elouga Bom, V. R. Bhardwaj, T. Ozaki, *Appl. Phys. Lett.* 98, 101104 (2011).
12. J.-F. Hergott, M. Kovacev, H. Merdji, C. Hubert, Y. Mairesse, E. Jean, P. Breger, P. Agostini, B. Carre, P. Salieres, *Phys. Rev. A* 66, 021801 (2002).
13. E. Takahashi, Y. Nabekawa, K. Midorikawa, *Opt. Lett.* 27, 1920 (2002).
14. T. D. Donnelly, T. Ditmire, K. Neuman, M. D. Perry, R. W. Falcone, *Phys. Rev. Lett.* 76, 2472 (1996).

15. L. B. Elouga Bom, R. A. Ganeev, J. Abdul-Hadi, F. Vidal, T. Ozaki, *Appl. Phys. Lett.* 94, 111108 (2009).
16. P. M. Paul, E. S. Toma, P. Breger, G. Mullot, F. Auge, P. Balcou, H. G. Muller, P. Agostini, *Science* 292, 1689 (2011).
17. A. Bahabad, M. M. Murnane, H. C. Kapteyn, *Nat. Photonics* 4, 570 (2010).
18. J. Seres, V. S. Yakovlev, E. Seres, C. H. Streli, P. Wobrauschek, C. H. Spielmann, F. Krausz, *Nat. Phys.* 3, 878 (2007).
19. A. H. Sheinfux, Z. Hems, M. Levin, A. Zigler, *Appl. Phys. Lett.* 98, 141110 (2011).
20. R. A. Ganeev, H. Singhal, P. A. Naik, U. Chakravarty, V. Arora, J. A. Chakera, R. A. Khan, M. Raghuramaiah, S. R. Kumbhare, R. P. Kushwaha, P. D. Gupta, *Appl. Phys. B* 87, 243 (2007).
21. R. A. Ganeev, C. Hutchison, T. Siegel, M. E. López-Arias, A. Zaïr, J. P. Marangos, *J. Modern Opt.* 58, 819 (2011).
22. R. A. Ganeev, C. Hutchison, T. Siegel, A. Zaïr, J. P. Marangos, *Phys. Rev. A* 83, 063837 (2011) .
23. R. A. Ganeev, C. Hutchison, A. Zaïr, T. Witting, F. Frank, W. A. Okell, J. W. G. Tisch, J. P. Marangos, *Opt. Express* 20, 90 (2012).
24. R. A. Ganeev, T. Witting, C. Hutchison, F. Frank, P. V. Redkin, W. A. Okell, D. Y. Lei, T. Roschuk, S. A. Maier, J. P. Marangos, J. W. G. Tisch, *Phys. Rev. A* 85, 015807 (2012).
25. W. Theobald, C Wulker, F. R. Schafer, B. N. Chichkov, *Opt. Commun.* 120, 177 (1995).
26. M. López-Arias, M. Oujja, M. Sanz, R. A. Ganeev, G. S. Boltaev, N. K. Satlikov, R. I. Tugushev, T. Usmanov, M. Castillejo, *J. Appl. Phys.* 111, 043111 (2012).
27. M. López-Arias, M. Oujja, M. Sanz, R. de Nalda, R. A. Ganeev, M. Castillejo, *Mol. Phys.* 110, 1651 (2012).
28. C. Hutchison, R. A. Ganeev, T. Witting, F. Frank, W. A. Okell, J. W. G. Tisch, J. P. Marangos, *Opt. Lett.* 37, 2064 (2012).
29. R. Torres, T. Siegel, L. Brugnera, I. Procino, J. G. Underwood, C. Altucci, R. Velotta, E. Springate, C. Froud, I. C. E. Turcu, M. Y. Ivanov, O. Smirnova, J. P. Marangos, *Opt. Express* 18, 3174 (2010).
30. C. Vozzi, M. Nisoli, J.-P. Caumes, G. Sansone, S. Stagira, S. De Silvestri, M. Vecchiocattivi, D. Bassi, M. Pascolini, L. Poletto, P. Villoresi, G. Tondello, *Appl. Phys. Lett.* 86, 111121 (2005).
31. J. Tate, T. Auguste, H. G. Muller, P. Salières, P. Agostini, L. F. DiMauro, *Phys. Rev. Lett.* 98, 013901 (2007).

32. K. Schiessl, K. L. Ishikawa, E. Persson, J. Burgdörfer, *Phys. Rev. Lett.* 99, 253903 (2007).

33. Y. Pertot, S. Chen, S. D. Khan, L. B. Elouga Bom, T. Ozaki, Z. Chang, *J. Phys. B: At. Mol. Opt. Phys.* 45, 074017 (2012).

34. R. A. Ganeev, C. Hutchison, T. Witting, F. Frank, W. A. Okell, A. Zaïr, S. Weber, P. V. Redkin, D. Y. Lei, T. Roschuk, S. A. Maier, I. López-Quintás, M. Martín, M. Castillejo, J. W. G. Tisch, J. P. Marangos, *J. Phys. B: At. Mol. Opt. Phys.* 45, 165402 (2012).

35. T. Witting, F. Frank, C. A. Arrell, W. A. Okell, J. P. Marangos, J. W. G. Tisch, *Opt. Lett.* 36, 1680 (2011).

36. R. A. Ganeev, V. V. Strelkov, C. Hutchison, A. Zaïr, D. Kilbane, M. A. Khokhlova, J. P. Marangos, *Phys. Rev.* 85, 023832 (2012).

37. J. S. Robinson, C. A. Haworth, H. Teng, R. A. Smith, J. P. Marangos, J. W. G. Tisch, *Appl. Phys. B* 85, 525 (2006).

38. M. Sanz, M. E. López-Arias, E. Rebollar, R. de Nalda, M. Castillejo, *J. Nanopart. Research* 13, 6621 (2011).

39. R. Torres, M. Jadraque, M. Martin, *Appl. Phys. A* 80, 1671 (2005).

40. M. Anselment, R. S. Smith, E. Daykin, L. F. Dimauro, *Chem. Phys. Lett.* 134, 444 (1987).

41. A. Rohlfing, *J. Chem. Phys.* 89, 6103 (1988).

42. S. Acquaviva, M. L. De Giorgi, *Appl. Surf. Sci.* 197–198, 21 (2002).

43. J. Roth, F. Géahler, H.-R. Trebin, *Int. J. Modern Phys. C* 11, 317 (2000).

44. T. E. Glover, *J. Opt. Soc. Am. B* 20, 125 (2003).

45. H.O. Jeschke, M.E. Garcia, K.H. Bennemann, *Phys. Rev. Lett.* 87, 015003 (2001).

46. A.V. Kabashin, M. Meunier, *J. Appl. Phys.* 94, 7941 (2003).

47. W. R. Creasy, J. T. Brenna, *Chem. Phys.* 126, 453 (1988).

48. S. C. O'Brien, J. R. Heath, R. F. Curl, R. E. Smalley, *J. Chem. Phys.* 88, 220 (1988).

49. J. W. G. Tisch, T. Ditmire, D. J. Frasery, N. Hay, M. B. Mason, E. Springate, J. P. Marangos, M. H. R. Hutchinson, *J. Phys. B At. Mol. Opt. Phys.* 30, L709 (1997).

50. H. Singhal, R. A. Ganeev, P. A. Naik, J. A. Chakera, U. Chakravarty, H. S. Vora, A. K. Srivastava, C. Mukherjee, C. P. Navathe, S. K. Deb, P.D. Gupta, *Phys. Rev. A* 82, 043821 (2010).

51. C. Altucci, R. Bruzzese, C. de Lisio, M. Nisoli, S. Stagira, S. De Silvestri, O. Svelto, A. Boscolo, P. Ceccherini, L. Poletto, G. Tondello, P. Villoresi, *Phys. Rev. A* 61, 021801R (2000).

52. C. A. Froud, E. T. F. Rogers, D. C. Hanna, W. S. Brocklesby, M. Praeger, A. M. de Paula, J. J. Baumberg, J. F. Frey, *Opt. Lett.* 31, 374 (2006).
53. S. Kim, J. Jin, Y.-J. Kim, I.-Y. Park, Y. Kim, S.-W. Kim, *Nature* 453, 757 (2008).
54. A. Husakou, J. Herrmann, *Phys. Rev. A* 83, 043839 (2011).
55. H. Kjeldsen, F. Folkmann, B. Kristensen, J. B. West, J. E. Hansen, *J. Phys. B: At. Mol. Opt. Phys.* 37. 1321 (2004).
56. G. V. Marr, J. B. West, *At. Data Nucl. Data Tabl.* 18, 497 (1976).
57. J. Levesque, D. Zeidler, J. P. Marangos, P. B. Corkum, D. M. Villeneuve, *Phys. Rev. Lett.* 98, 183903 (2007).
58. A. D. Shiner, B. E. Schmidt, C. Trallero-Herrero, H. J. Wörner, S. Patchkovskii, P. B. Corkum, J.-C. Kieffer, F. Légaré, D. M. Villeneuve, *Nature Phys.* 7, 464 (2011).
59. R. A. Ganeev, M. Suzuki, T. Ozaki, M. Baba, H. Kuroda, *Opt. Lett.* 31, 1699 (2006).
60. R. A. Ganeev, P. A. Naik, H. Singhal, J. A. Chakera, P. D. Gupta, *Opt. Lett.* 32, 65 (2007).
61. R. A. Ganeev, L. B. Elouga Bom, J.-C. Kieffer, T. Ozaki, *Phys. Rev. A* 76, 023831 (2007).
62. R. A. Ganeev, M. Suzuki, M. Baba, H. Kuroda, *Appl. Phys. Lett.* 94, 051101 (2009).
63. M. Hentschel, R. Kienberger, C. Spielmann, G. A. Reider, N. Milošević, T. Brabec, P. Corkum, U. Heinzmann, M. Drescher, F. Krausz, *Nature* 414, 509 (2001).
64. T. Witting, F. Frank, W. A. Okell, C. A. Arrell, J. P. Marangos, J. W. G. Tisch, *J. Phys. B: At. Mol. Opt. Phys.* 45, 074014 (2012).
65. C. Altucci, J. W. G. Tisch, R. Velotta, *J. Modern Opt.* 58, 1585 (2011).
66. R. A. Ganeev, T. Witting, C. Hutchison, F. Frank, M. Tudorovskaya, M. Lein, W. A. Okell, A. Zaïr, J. P. Marangos, J. W. G. Tisch, *Opt. Express* 20, 25239 (2012).
67. D. Kilbane, E. T. Kennedy, J. P. Mosnier, J. T. Costello, P. van Kampen, *J. Phys. B: At. Mol. Opt. Phys.* 38, 83 (2005).
68. V. K. Dolmatov, *J. Phys. B: At. Mol. Opt. Phys.* 29, L687 (1996).
69. J. B. West, J. E. Hansen, B. Kristensen, F. Folkmann, H. Kjeldsen, *J. Phys. B: At. Mol. Opt. Phys.* 36, L327 (2003).
70. P. V. Redkin, R. A. Ganeev, *Phys. Rev. A* 81, 063825 (2010).
71. V. Strelkov, *Phys. Rev. Lett.* 104, 123901 (2010).

72. M. V. Frolov, N. L. Manakov, A. F. Starace, *Phys. Rev. A* 82, 023424 (2010).
73. M. Tudorovskaya, M. Lein, *Phys. Rev., Ser. A* 84, 013430 (2011).
74. J. Yao, B. Zeng, W. Chu, J. Ni, Y. Cheng, *J. Modern Opt.* 59, 245 (2012).

Chapter 6

Characterization of Plasma Harmonics

In this chapter, I show some peculiarities in characterization of plasma harmonics. They are mainly centered in definition of new features of this process. Among them, the following results will be discussed:

(a) I show the results of theoretical and experimental studies of HHG in the ensemble of silver atoms irradiated by intense femtosecond pulses of Ti:sapphire laser. It is shown during these studies that the photoemission spectra exhibit unusual behavior when the laser field strength approaches near-atomic values. In subatomic field strength, the cutoff frequency (CF) increases linearly with laser pulse intensity. However, when the field strength approaches near-atomic region, firstly CF slows down and then saturates. To give new interpretation of such kind of photoemission spectrum behavior, the light-atom interaction theory based on the use of eigenfunctions of boundary value problem for "the atom in an external field" was proposed instead of the traditional basis of the "free atom" eigenfunctions. The results of computer simulations clearly demonstrate saturation of the CF at near-atomic strength of a laser field. The problem of angular divergence of a harmonic emission is analyzed.

Plasma Harmonics
Rashid A. Ganeev

ISBN 978-981-4463-30-0 (Hardcover), 978-981-4463-31-7 (eBook)
www.panstanford.com

The comparison of experiment and theory shows a reasonable coincidence.

(b) I analyze the laser-induced plasma formation for harmonics generation through the calculations of particle concentration in the interaction area in the cases of pulses of different duration. It has been found that high particle concentration leads to formation of nanoparticles, which enhances the harmonic generation efficiency from laser plasmas relative to the gas jets at the same conditions.

(c) I present the studies of HHG in laser ablation plumes of the ribonucleic acid nucleobase uracil and its deoxyribonucleic acid counterpart thymine. Harmonics were generated using 780 nm, 30 fs and 1300 nm, 40 fs radiation upon ablation with 1064 nm, 10 ns or 780 nm, 160 ps pulses. Strong HHG signals were observed from uracil plumes with harmonics emitted with photon energies greater than 55 eV. Results obtained in uracil plumes were compared with those from thymine, which did not yield signs of harmonic generation. The ablation plumes of the two compounds were examined by collection of the ablation debris on a silicon substrate placed in close proximity to the target and by time-of-flight mass spectrometry. From these evidences, we conclude that the differences in HHG signal are due to the different fragmentation dynamics of the molecules in the plasma plume. These studies constitute the first attempt to analyze differences in structural properties of complex molecules through plasma ablation induced HHG spectroscopy.

6.1 High-Order Harmonic Cutoff Frequency in Atomic Silver Irradiated by Femtosecond Laser Pulses: Theory and Experiment

In this section, the saturation of the CF of the photoemission spectra generated in silver by a femtosecond laser pulse is discussed [1]. In experiments with 150 fs pulse of Ti:sapphire laser, the highest experimentally measured harmonic is 61 and the saturation intensity is $I_s^{(e)} = (4 \pm 1) \times 10^{14}$ W/cm^2. The theoretical analysis is based on the non-perturbative theory of nonlinear atomic response, which enables us to vary the laser field strength in wide region from a subatomic to over-atomic values. The results of computer simulations have shown that the cutoff is also saturated, but the maximal harmonic order exceeds 300 in the case of a single atom. This drastic difference is due to the dispersion effects, which play the decisive role in the spatially distributed ensemble of ionized atoms. At given experimental conditions, the phase mismatch between the driving laser pulse and the harmonic emission modifies significantly the frequency spectra of an atomic ensemble in comparison with the single atom response. The amplitudes of the high harmonics are significantly suppressed. The simplest way to avoid this modification is to decrease the atomic density. However, this way is in contradiction with the perspectives of practical applications of the high-order harmonic emission. Nevertheless, the development of methods for control of the high harmonic emission spectra is one of the most important problems in modern nonlinear optics.

The benefits of combined experimental and theoretical study of the CF saturation effect are stated as follows. The coincidence of the saturation intensity provides us the basis to link the CF saturation with the total atomic ionization effect. At the same time, the results of numerical calculations indicate clearly the optimal laser pulse intensity at which the CF in a single atom emission is a maximum. Further, along with the modification of frequency response spectra, the dispersion effects strongly influence the angular spectra of an atomic ensemble emission. The computer modeling provides the possibility to estimate the material parameters of medium, such as as atomic, electron, and ion density, as well as the characteristic geometrical sizes of interaction area.

6.1.1 *Description of the Problem*

In spite of the 20-year history, the effect of the HHG is still under a great interest of both experimentalists and theoreticians. The origin of this interest is manifold. From a practical point of view, the HHG is one of the effective mechanisms for producing a coherent emission in a broad region of electromagnetic wave spectrum. The presence of plateau region in the harmonic amplitude distribution in XUV region affords grounds for development of subfemtosecond pulse formation methods. As a result, the new frontiers are opened up in science by extending the nonlinear optics and time-resolved spectroscopy to the XUV region [2] and pushing ultrafast science to the attosecond domain, enabling the XUV spectroscopy and imaging of molecular orbitals [3], surface dynamics [4], and electron motion. The HHG is the reliable route to produce attosecond light pulses [5] and is therefore fundamental to attosecond science [6].

At present, the efficiency of conversion to high-order harmonics is really too small to consider this emission as a real coherent XUV source for biology, plasma diagnostics, medicine, microscopy, photolithography, and so on. Hence, the search for ways of increasing the CF and the HHG efficiency in the XUV spectral range is still among the most topical problems of nonlinear optics.

The HHG effect was observed with the large number of periodic table elements usually having small and middle atomic numbers [7–11]. As a rule, the interaction medium is cell, gas jet, or laser plasma that is prepared by irradiation of metal surface by laser heating pulse. The maximum harmonic order, or the CF, obtained in plasma media to date varies from the sixties to seventies harmonics of fundamental frequency [7, 12] of Ti:sapphire laser. The highest-order harmonics (the 101st harmonic, $\lambda = 7.9$ nm) have been obtained in manganese plasmas [8]. The efficiency of conversion in the plateau region amounts to nearly 10^{-5} [9]. This value depends on both atomic energy level diagram and laser pulse parameters (intensity, energy and duration, envelope temporal profile, carrier envelope phase). The CF movement into the sub-nanometer region promises new possibilities in creating x-ray coherent sources. Hence, the study of the CF dependence on the irradiated media properties and external laser pulse parameters is the problem of significant scientific and technical interest. The CF values corresponding to seventies harmonics have been obtained in experiments with silver plasma [10] and this stimulated us to choose this medium for numerical simulations and compare the obtained results with the experimental data.

Up to now, a number of different theoretical models to describe the dynamics of an atomic electron in a strong laser field have been developed. These models are based on different approximations, which are usually valid in the restricted area of laser pulse field strength. However, the unified theory applicable for both weak and strong laser fields has not been yet developed. Indeed, let us make some short review of the proposed theoretical approaches. The foundation of the atom ionization theory has been laid by the paper of Keldysh [13]. The equation for an ionization rate proposed by Keldysh is the most widely used formula in the various study of ionization process. The Keldysh theory has got further development in the series of papers [14–16]. The theory developed in these papers is known as the Perelomov–Popov–Terent'ev (PPT) model. The modification of these approaches, which has been proposed in [17], is usually called by Keldysh–Faisal–Reiss (KFR) approximation. The model known as the strong-field approximation (SFA) was proposed and developed in the series of papers by Reiss [18–20]. In contrast to the previous theories, this approximation does not use the saddle point method in calculating the matrix element of the transition from initial atomic state to ionized state of the Volkov continuum and, hence, it does not need the assumption that the photon energy is much smaller than electron binding energy. The Kramers–Henneberger model developed in [21–26] is the model of a dressed atom, the ionization potential of which decreases with laser field strengthening.

Many important results in the theory of harmonic generation have been obtained with the use of the zero range pseudopotential model [27]. This model, originally formulated for the case of a linearly polarized laser field, has been later generalized for the cases of a one-color [28] and a two-color [29] fields with arbitrary polarization. Classical theory of the HHG is based on the classical analysis of the problem on an atomic electron motion in the external laser field. The classical Newton equations are solved here for an ensemble of trajectories generated from an initial electron distribution in the phase space. This distribution is supposed to mimic the true quantum initial state of the system, so the averages over this distribution are analogs of the quantum averages [30, 31].

The method of direct numerical solution of the time-dependent Schrödinger equation (TDSE) became widely used in computer simulations [32, 33]. In the 1D case, the direct integration of the TDSE can be performed with the either finite element

(Crank–Nicholson), or the split operator techniques; in the context of harmonic generation, this method has been firstly used in [34]. The development of numerical methods for the 3D case has been initiated in [35, 36], where the 2D finite element ("grid") method was used. Later, it has been shown that the basis expansion methods, which employ the symmetry of the problem, work much better [37–40]. The modern codes are usually based on angular momentum basis expansion with a subsequent solving of the coupled set of equations for the radial wave functions with the help of the finite grid methods [41]; the Sturmian expansion [42, 43] or the B-spline expansion [44, 45] is also used. Most of those codes are quite powerful and allow calculating the atomic response directly without adiabatic approximation. Unfortunately, there are some disadvantages of these methods. Firstly, they are quite time-consuming. Secondly, any computer simulation inevitably deals with a modeled atom. To avoid the singularity of the Coulomb potential, some empirical approximations for intra-atomic potential are usually used (see e.g., [26] and references therein). Thirdly, it is often supposed that the wave functions of the continuum spectrum states are the Volkov wavefunctions. However, in the limit of zero field strength, these wave functions tend to the free particle wave functions but not to the wave functions of a particle in the Coulomb field. The problem of gauge choice in the numerical modeling has been discussed in [46].

The traditional interpretation of the harmonic generation is usually based on the model of three-stage process that comprises the ionization of an atom, the electron acceleration in the electromagnetic field, and its subsequent collision with ion (see Chapter 1). This process is periodically repeated every half cycle of the electromagnetic wave. The evolution of the ionized electron is usually described with the help of Volkov wave functions or classical mechanics equations. It should be noted that the use of the Volkov wave functions is due to the temporal evolution of the continuous spectrum electron wave functions in the presence of the laser field. At the same time, the spatial profile of the discrete spectrum wave functions is also changed in the presence of the external electromagnetic field. However, in the limit of zero strength fields, the Volkov wave functions are transformed to the free electron wave functions, while the discrete spectrum wave functions became the atomic eigenfunctions (e.g., hydrogenic ones). Hence, in the weak or zero fields, the eigenfunctions of discrete and continuous spectra are not mutually orthogonal and they do not compose the complete basis of the orthogonal functions. This is the most principle inconsistence of the above methods.

The self-consistent theory of the HHG based on the analytic solution of the TDSE has been presented in [47]. This theory is of great interest because it gives simple analytic formulas for the CF of photoemission spectra as a function of laser field parameters. It enables to analyze the harmonic generation efficiency as a function of parameters of laser pulse and atomic energy-level structure. However, the parameters of laser fields in the modern experiments lie in the region that is far from the main assumptions of this theory. Firstly, the spectral width of the femtosecond pulses, which are widely used now, exceeds significantly the main constraints of considered theory. Secondly, the usage of dipole approximation for interaction Hamiltonian results in evident restrictions in laser field intensity, because the laser field strength should be much smaller than the intra-atomic field strength. Finally, the intense femtosecond laser pulses populate effectively the excited states of the discrete atomic spectrum [48, 49], therefore if we neglect the population of the excited states, then we shall get the misinterpretation of the experimental dependences. Of course, the excited states can be easily included in frames of approach developed in [47].

A number of different non-perturbative approaches to the light-atom interaction problem have been developed recently (see, e.g., [50–52]). Most of them are based on the Hamiltonian in the dipole approximation with the use ***Ed***-gauge [50, 51] or ***Ap***-gauge [52]. Here, we develop an alternative approach to the theory of a light-atom interaction, the idea of which has been initially proposed in [53]. It is shown that the boundary value problem for an electron moving in the superposition of a centro-symmetric intra-atomic field and a field of an external electromagnetic wave has exact mathematical solutions in the non-relativistic approximation. The spectrum of eigenvalues for this problem coincides with that for the free atom boundary value problem and eigenfunctions $\varphi_n(\mathbf{r}, t)$ are related with the free atom eigenfuctions $u_n(\mathbf{r})$ by $\varphi_n(\mathbf{r}, t) =$

$u_n(\boldsymbol{r}) \exp(ie\boldsymbol{A}(t) r/\hbar c)$ [49, 53–56]. The set of eigenfunctions $\varphi_n(\boldsymbol{r}, t)$ provides the complete basis of the orthonormal functions, which is used to calculate the matrix elements of quantum-mechanical operators. In the case of a hydrogen-like atom or ion, the dynamics of an atomic electron evolution over the states of both discrete and continuous spectra is calculated in consistent mathematical form. The main advantage of the proposed approach is its non-perturbative manner. The ratio of laser field strength E_0 to intra-atomic field strength, $E_{at} = e/a_B^2$ (where a_B is the Bohr radius), should not be the smallness parameter of the problem and it could take the arbitrary values in the non-relativistic region. This approach has been exploited in [49] to study the specific features of energy spectra of photoelectrons arising in the process of silver atom interaction with femtosecond laser pulses.

In this section, we analyze single silver atom photoemission spectra. The most unusual feature of the single atom photoemission spectra is the CF saturation effect. The CF saturation means that the maximum observed harmonic order ceases to be intensity dependent at $I \geq I_{sat}$. For the first time, the effect of the CF saturation in silver plasma was experimentally observed in [10]. The maximal order of experimentally observed harmonic was equal to 71. On the contrary, the results of computer simulations performed for the single silver atom [54] have shown that the CF is saturated in the region of overatomic field strength, but at near atomic field strength, the maximal order of generated harmonics is much higher than that observed experimentally. The main purpose of the present study is to explain the difference between the experimental and theoretical results and to provide the interpretation of the main mechanisms lying in the origin of the CF saturation effect.

6.1.2 *Experimental Studies*

To study the high-order nonlinearities through the HHG, the silver target was placed inside the vacuum chamber to ignite the plume by laser ablation of this sample. Silver has previously proven to be most efficient medium for production of the plasma plumes where the highest HHG efficiency (8×10^{-6}) in the XUV spectral range was observed [57].

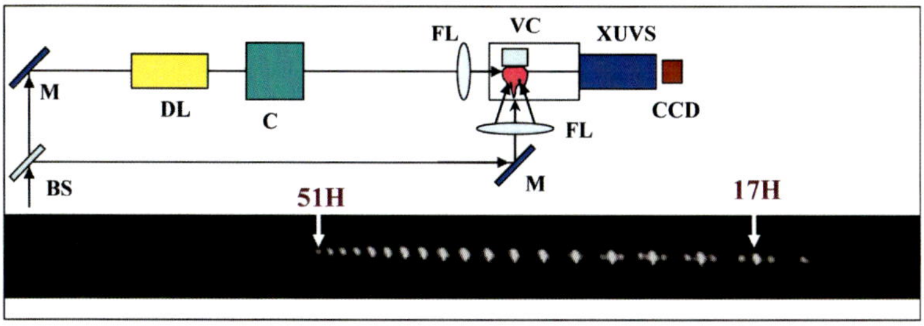

Figure 6.1 Experimental setup for HHG in silver plasma. BS, beam splitter; M, mirrors; DL, delay line; C, compressor, FL, focusing lenses; VC, vacuum chamber; XUVS, extreme ultraviolet spectrometer; CCD, charge-coupled device. Inset: Typical image of harmonic distribution from 17th to 51st orders. Reproduced from Ref. 1 with kind permission from The European Physical Journal.

The pump source was a commercial, chirped pulse amplification Ti:sapphire laser, whose output was further amplified using a homemade three-pass amplifier operated at a 10 Hz pulse repetition rate. A heating pulse was split from the amplified laser beam by a beam splitter before a pulse compressor (Fig. 6.1). The heating pulse duration was 210 ps. A spherical lens focused the heating pulse on the target placed in the vacuum chamber. The area of ablation was equal to 0.5 mm^2. The 210 ps heating pulse intensity at the target surface was varied in the range of $I_{pp} = (0.5–5) \times 10^{10}$ W/cm^2 by changing the energy of heating pulse. A driving femtosecond pulse at a central wavelength of 792 nm had the energy of 12 mJ and pulse duration of 150 fs after the propagation of the compressor stage. After the proper delay with regard to the heating pulse irradiation (20–80 ns), the driving pulse was focused by a spherical lens ($f/10$) on the ablation plume from the orthogonal direction. The maximum intensity of the driving pulse at the focal spot could reach 6×10^{16} W/cm^2. The position of laser focus was adjusted to be either before or after the laser plume to exclude the influence of considerable ionization of the silver neutrals and ions in laser plume and to maximize the harmonic yield. The intensity of the driving pulse (I_{fp}) at the plasma area was adjusted between 1×10^{14} and 1×10^{15} W/cm^2. The distance between the HHG interaction area and target surface was varied in the range of 100–200 μm. The spectrum

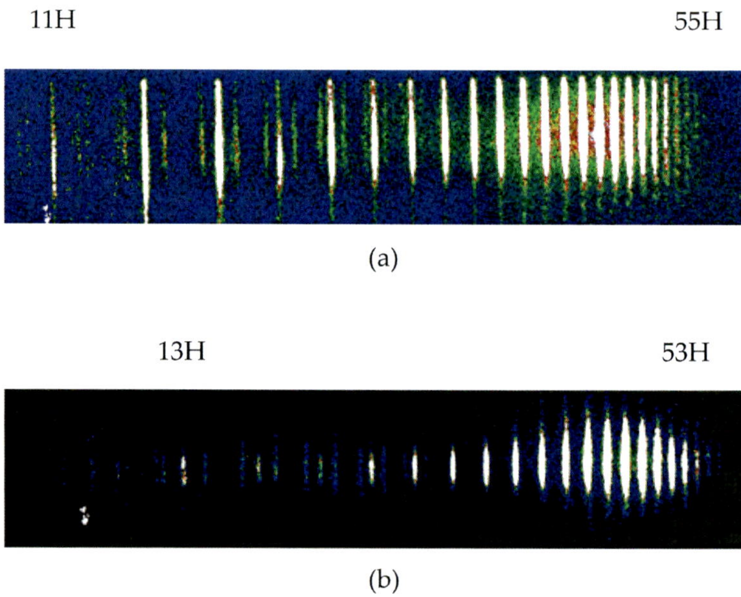

Figure 6.2 (a) Harmonic spectra from silver plasma in the case of focusing behind the plasma area. (b) Spectral image of harmonics from silver plasma in the case of aperturing the input femtosecond radiation. Reproduced from Ref. 1 with kind permission from The European Physical Journal.

of generated high-order harmonics was analyzed by a grazing incidence XUV spectrometer with a gold-coated flat-field grating. The XUV spectrum was detected using a microchannel plate with a phosphor screen readout, and the optical output from the phosphor screen was recorded using a charge-coupled device camera.

Figure 6.2a shows the image of harmonic spectrum from Ag plasma at the conditions when the femtosecond pulse was focused after the plasma plume. The harmonics up to the 51st order were routinely observed at these conditions, when the intensity of femtosecond pulse at the plasma area was $I_{fs} = 2.5 \times 10^{14}$ W/cm^2. Note that divergences of harmonics at the beginning of plateau-like distribution of harmonics and at the cutoff region were approximately equal to each other. The increase of laser intensity at the plasma area led to increase of harmonic cutoff (H_c) up to the 61st order. The divergence of harmonics gradually increased with the growth of harmonic order and the divergences of the lowest (17th–25th) were approximately two to three times less compared with the highest (45th–57th) harmonics.

At these experimental conditions (focusing inside the plasma plume, intensity of the driving pulse at the plasma area $I_{fs} = 4 \times 10^{14}$ W/cm^2, delay between the pulses of 30 ns), the divergence of driving radiation was changed by introducing the aperture in front of the input window of vacuum chamber through which femtosecond radiation propagated toward the plasma plume. A decrease of beam sizes led to considerable deformation of observed spectra, when one can find the conditions of variations of the divergence of harmonics in different ranges of harmonic spectra, as well as considerable redistribution of harmonic efficiency along the plateau region. When we decreased the aperture down to 4 mm (and correspondingly decreased the energy of propagated radiation from 12 mJ to 4 mJ), a remarkable pattern of harmonic spectra appeared showing the five to 10 times enhancement of harmonic intensity along the axis of laser propagation for higher harmonics compared with lower ones (Fig. 6.2b). The on-axis intensities of 11th–19th harmonics were considerably less than those in the range of 25th–47th orders.

The $H_c(I_{fp})$ dependence becomes saturated at $I_{fp} \approx (4 \pm 1) \times 10^{14}$ W/cm^2 (see Fig. 6.3 showing the harmonic cutoff dependence between the maximum observed harmonic order and the femtosecond pulse intensity). The intensity $I_{fp} = 3.5 \times 10^{14}$ W/cm^2 is well above the one at which the neutral Ag atoms are ionized. So, it may be an additional indication of the crucial role of singly charged ions in generation of high-order harmonics.

The novelties of these experiments are (1) the observation of the efficient harmonics generation for the high orders, while keeping the low-order ones at smaller conversion efficiency, and (2) the observations of saturation of HHG above some level of laser intensity [$(4 \pm 1) \times 10^{14}$ W/cm^2)]. The aperturing of laser beam led to better phase-matching conditions for higher orders compared with lowest harmonics (Fig. 6.2b). The intensity of those high-order harmonics became strong enough to allow the observations of second-order diffraction patterns from the grating (see the lobes from the left and right sides of lowest harmonics on Fig. 6.2a).

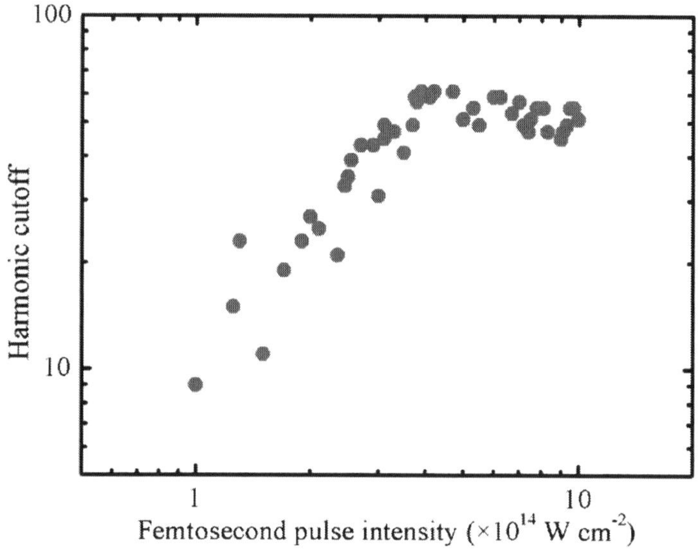

Figure 6.3 Harmonic cutoff from silver plasma as a function of laser intensity. The saturation of $H_c(I_{fp})$ dependence was started at $(4\pm1) \times 10^{14}$ W/cm^2. Reproduced from Ref. 1 with kind permission from The European Physical Journal.

The violation of the phase-matching conditions at high intensities of driving radiation is caused by the appearance of additional free electrons possessing positive dispersion, which does not match with the relatively low dispersion of atoms and ions. The difference with the case of analogous observations in the gaseous media is related with the presence of the free electrons at the initial state of plasma media. The ablation at appropriate conditions can create the formation of "optimal" plasma, which, although containing free electrons, does not prevent the violation of phase matching. Only additional free electrons appearing during further ionization of singly charged ions of silver (second ionization potential of silver is 21 eV, which corresponds to the breakdown saturation intensity of laser radiation of $\sim 3 \times 10^{14}$ W/cm^2).

Silver is one of most attractive targets for plasma formation and HHG. Conversion efficiency reported for this plasma medium for the harmonics in the range of photon energies of 50–80 eV (8×10^{-6}) [10] is the highest among other plasma plumes. Moreover, the value of conversion efficiency is the same, as those reported in the case of gas HHG studies. Assuming the stability amendments in the harmonic generation from this plasma (of the order of 1 million shots at 1 kHz pulse repetition rate [59]) one can anticipate a robust application of this target for various applications of plasma HHG spectroscopy.

6.1.3 *Theory*

The process of a light-atom interaction became fundamentally the non-resonant process, when the laser pulse temporal width approaches to a few cycles of optical oscillations. Indeed, in this case, the pulse spectral width is comparable or exceeds the energy-level difference between the free atom states. Hence, the well developed theories of the resonant one-photon or multi-photon interactions cease to be valid in this case. Moreover, the binding energy of atomic electrons tends to zero when the laser field strength exceeds the intra-atomic field strength, because the electron moves mainly in the spatial regions, which are far from its parent nuclei. It is evident that the basis of a free atom eigenstates could not serve as a useful basis for the wave function expansion when the laser field strength approaches to the intra-atomic values. At the same time, it has been shown in [53] that there is a direct mathematical relationship between the eigensolutions of the "free atom" boundary value problem and boundary value problem for "the atom in an external field." Let us now turn on to the overview of basic principles lying in the ground of proposed mathematical formalism.

In the non-relativistic case, the theory of light-atom interaction is based on the TDSE:

$$i\hbar\frac{\partial\psi}{\partial t} = \left[\frac{1}{2m}\left(\boldsymbol{p} - \frac{q}{c}\boldsymbol{A}\left(t\right)\right)^{2} + U\left(\boldsymbol{r}\right)\right]\psi, \qquad (6.1)$$

where $U(\boldsymbol{r})$ is the potential energy of electron in the intra-atomic field and $\boldsymbol{A}(t)$ is the vector potential of the external electromagnetic wave. The traditional approach to the mathematical analysis of Eq. (6.1) is based on the expansion of wave functions of the TDSE into the series of the "free atom" eigenfunctions, which are the eigensolutions of the following boundary value problem

$$\left(\frac{p^2}{2m} + U(\mathbf{r})\right) u_{\mathrm{N}}(\mathbf{r}) = E_{\mathrm{N}} u_{\mathrm{N}}(\mathbf{r}). \tag{6.2}$$

As far as the boundary value problem is a three-dimensional one, then the quantum number N is really a set of three quantum numbers. For example, in the case of a centrosymmetric intra-atomic field, $U(r) = U(\mathbf{r})$, the quantum number N is the set of the following three quantum numbers: $N = (n, l, m)$, where n is the principal (or radial) quantum number, l is the angular momentum, and m is the projection of angular momentum. The eigenfunctions of the boundary value problem (6.2) for any atom with a centrosymmetric intra-atomic field read as

$$u_{\mathrm{N}}(\mathbf{r}) = R_{\mathrm{nl}}(\mathbf{r})\, Y_{\mathrm{lm}}(\theta, \varphi), \tag{6.3}$$

where $Y_{\mathrm{lm}}(\theta, \varphi)$ are the spherical harmonics and $R_{nl}(r)$ are the radial parts of wave functions, which should satisfy the boundary conditions at points $r = 0$ and $r = \infty$.

Let us now turn to the following boundary value problem

$$\left[\frac{1}{2m}\left(\vec{p} - \frac{q}{c}\chi(\mathbf{r}, t)\right)^2 + U(\mathbf{r})\right] \varphi_{\mathrm{M}}(\mathbf{r}, t) = E_{\mathrm{M}} \varphi_{\mathrm{M}}(\mathbf{r}, t), \tag{6.4}$$

where $\chi(r, t)$ is an arbitrary function of space-time coordinates. The eigensolutions of the boundary value problem (6.4) are

$$\varphi_{\mathrm{M}}(r, t) = u_{\mathrm{N}}(r) \exp\left(i \frac{q}{\hbar c} \chi(\mathbf{r}, t)\right), \tag{6.5}$$

$$E_{\mathrm{M}} = E_{\mathrm{N}},$$

where $u_{\mathrm{N}}(\mathbf{r})$ is the eigenfunction of the problem (6.2), which corresponds to the eigenvalue E_{N} of this problem. Indeed, substituting (6.5) into (6.4), one gets Eq. (6.2). So, the eigenvalues of the problems (6.2) and (6.4) coincide exactly, while the eigenfunctions are different.

Let the function $\chi(\mathbf{r}, t)$ be $\chi(\mathbf{r}, t) = \mathbf{A}(t)\,\mathbf{r}$, then the operator in the left-hand side of the boundary value problem (6.4) coincides exactly with the Hamiltonian of Eq. (6.1). From the general point of view, the variable t is some external parameter of the boundary value problem (6.4). The variation of this parameter results in variation of the spatial profile of eigenfunctions $\varphi_{\mathrm{M}}(\mathbf{r}, t)$, but it does not affect on the magnitude of eigenvalue E_{M}. So, one can see that, in spite of the electron quiver motion in the laser field, it can still be in the state of the same energy, $E_{\mathrm{M}} = E_{\mathrm{N}}$; at the same time, the spatial profile of electron wave function varies in time (for details please see [56]).

Notice that the symmetry properties of the Hamiltonians of the problems (6.2) and (6.4) are different in general case, because the symmetry of the Hamiltonian (6.4) depends on the explicit form of the function $\chi(\mathbf{r}, t)$. As a result, the symmetry properties of the eigenfunctions of these two problems are also different. According to the well known theorems of mathematical physics, the eigenfunctions of the boundary value problem (6.2) constitute the complete orthonormalized basis of functions. At the same time, one can easily see that

$$\int \varphi_n^*(r,t)\,\varphi_m(r,t)\,dV = \int u_n^*(r)\,u_m(r)\,dV = \delta_{nm},$$

$$\sum_n \varphi_n^*(r,t)\,\varphi_n(r',t) = exp\left[-i\frac{q}{\hbar c}\left(\chi(r,t) - \chi(r',t)\right)\right] \times \sum_n u_n^*(r)\,u_n(r') = \delta(r - r').$$

Hence, the set of the eigenfunctions (6.5) also forms the complete orthonormalized basis. Thus, any eigenfunction of one basis can be expressed in terms of another one

$$\varphi_n(r,t) = \sum_m C_{nm}(t)\,u_m(r), \tag{6.6}$$

$$u_n(r) = \sum_m D_{nm}(t)\,\varphi_m(r,t), \tag{6.7}$$

where coefficients of expansions are defined by

$$C_{nm}(t) = \int u_m^*(r)\exp\left(i\frac{q}{\hbar c}\chi(r,t)\right)u_n(\mathbf{r})\,dV,$$

$$D_{nm}(t) = \int \varphi_m^*(r,t)\,u_n(\mathbf{r})\,dV = \int u_m^*(r)\exp\left(-i\frac{q}{\hbar c}\chi(r,t)\right)u_n(\mathbf{r})\,dV.$$

Substituting (6.6) into (6.7), we get

$$\sum_m C_{nm}D_{mp} = \delta_{np}.$$

It is convenient to introduce the following transformation matrix

$$V_{nm}(t) = \int u_n^*(\mathbf{r}) \exp\left(-i\frac{q}{\hbar c}\chi(\mathbf{r}, t)\right) u_m(\mathbf{r})\, dV. \tag{6.8}$$

As we have mentioned above, in the case

$$\chi(\mathbf{r}, t) = \mathbf{A}(t)\,\mathbf{r}, \tag{6.9}$$

the Hamiltonian of the problem (6.4) takes the form

$$H = \frac{1}{2m}\left(\mathbf{p} - \frac{q}{c}\mathbf{A}(t)\right)^2 + U(r).$$

and coincides with the Hamiltonian of Eq. (6.1). Hence, the eigenfunctons of the Hamiltonian of the TDSE are

$$\varphi_n(\mathbf{r}, t) = u_n(\mathbf{r}) \exp\left(i\frac{q}{\hbar c}\mathbf{A}(t)\,\mathbf{r}\right). \tag{6.10}$$

One can see that Eq. (6.10) looks similar to the formula for gauge transformation proposed by Goppert–Mayer [58]. However, it should be noted here that this is only the formal resemblance, because the wave functions (6.10) are really the eigenfunctions of the boundary value problem (6.4).

At the same time, the bases of the eigenfunctions of the boundary value problems (6.2) and (6.4) are related one-to-one by transformations (6.6)–(6.7). So, we can expand the wavefunction of the TDSE (1) into series of free atom eigenfunctions (6.3)

$$\psi = \sum_{nlm} a_{nlm}(t)\, u_{nlm}(\mathbf{r}) \tag{6.11}$$

and then use the relations (6.7) to calculate the matrix elements of operators. Note that in (6.11) we have used the three-dimensional indexes of atomic states to remind that summation over quantum number n implies the summation over discrete spectrum states and integration over continuum spectrum states.

Executing some evident transformations, we get the following set of differential equations for probability amplitudes of discrete and continuum spectrum states.

$$i\hbar\frac{da_n}{dt} = \sum_{m,k} V_{nk}^{-1} E_k V_{km} a_m. \tag{6.12}$$

It is well known that the number of discrete states of any hydrogen-like atom is infinite and there is an innumerable number of continuum spectrum states. Thus, the accuracy of numerical calculations depends substantially on the selection of discrete spectrum states and magnitude of discretization step for states of continuum spectrum. It is easy to see from Eq. (6.12) that the dynamics of population amplitudes depends significantly on the matrix elements $V_{nm}(t)$. It has been shown in Refs. [49] and [53–55] that in the case of a hydrogen-like atom, these matrix elements can be calculated in an explicit analytic form. Hence, in the frame of developed approach, the rates of transitions between the different atomic states are calculated in the explicit analytic form. What is most important that the developed approach provides the consistent mathematical procedure of a selection of atomic levels playing the main role in evolution of a state of the atom in an external laser field. Indeed, according to the orthogonality condition, we get

$$\int \varphi_n^*(\mathbf{r}, t)\,\varphi_m(\mathbf{r}, t)\,dV = \sum_{p=1}^{\infty} V_{pn} V_{pm}^* = \delta_{nm}$$

Specifically, the expression

$$\int |\varphi_n(\mathbf{r}, t)|^2\,dV = \sum_{m=1}^{\infty} |V_{mn}|^2 = 1$$

identifies the completeness of the expansion of the wave function φ_n into the series of eigenfunctions u_m. Each term $|V_{mn}|^2$ determines the relative contribution of the state $u_m(r)$ in this expansion.

Let us introduce the following sum

$$S_N^{(n)} = \sum_{m=1}^{N} |V_{mn}|^2 . \tag{6.13}$$

If we take into account only some finite number of discrete and continuum states, then the sum (6.13) became the function of the field strength E_0/E_{at}, and it could not be identically equal to unity in whole range of the field amplitude variation. But if we make only summation over the states that we have taken into account, then the magnitude of this sum will serve as a measure of the expansion completeness. Figure 6.4 shows the dependence of the sum $S_N^{(1)}$ on the dimensionless field strength μ_0, for the cases when the following states are taken into account: the hydrogen ground state (1*s*) only (squares), five low-lying excited states and the ground state

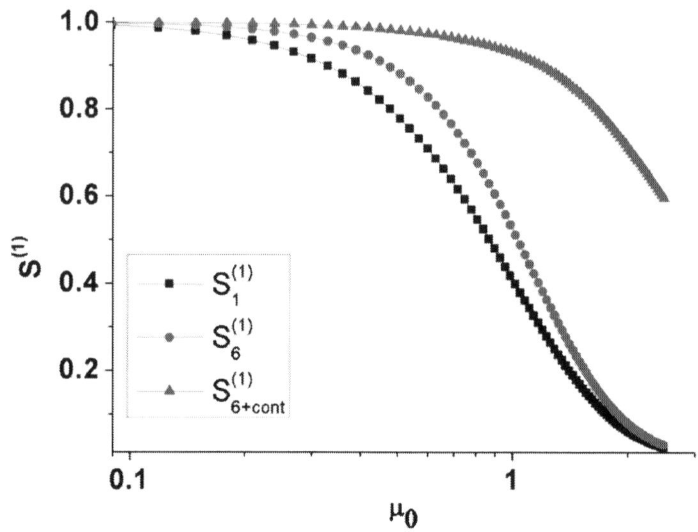

Figure 6.4 The sum $S_N^{(n)}$ as a function of parameter μ_0 for the different finite sets of a hydrogen atom eigenfunctions. Reproduced from Ref. 1 with kind permission from The European Physical Journal.

(circles), and with addition of the states of continuum band with orbital quantum numbers $l = 0, 1, 2$ (triangles). The field strength parameter μ_0 is defined in the next subsection (see Eq. 6.15). Notice, that in calculating of the $S_N^{(1)}$ we have made an integration over all energy states of continuum band. Figure 6.4 demonstrates the convergence of mutual expansion of eigenfunctions of the problems (6.2) and (6.4).

In computer simulations, we shall assume that a silver atom interacts with the linear polarized laser pulse of the Gaussian temporal profile

$$E = E_0 \exp\left(-\frac{(t-t_0)^2}{\tau_0^2}\right)\sin(\omega_0 t), \tag{6.14}$$

where ω_0 is the pulse carrier frequency and τ_0 is the pulse duration. The delay time t_0 is chosen to fit the pulse temporal profile with the center of an integration interval.

It can be seen from Eqs. (6.8) and (6.10) that the governing parameter of the problem is

$$\mu_0 = \frac{eA_0a_0}{\hbar c} = \frac{eE_0a_0}{\hbar\omega_0}, \tag{6.15}$$

where A_0 is the amplitude of the laser filed vector potential and a_0 is the radius of the valence electron orbit. In the case of a hydrogen atom, the radius of orbit is the Bohr radius: $a_0 = a_B$, and the latter equation became

$$\mu_0 = \frac{eA_0 a_B}{\hbar c} = \frac{eE_0 a_B}{\hbar \omega_0} = \frac{2U_0}{\hbar \omega_0} \frac{E_0}{E_{at}},$$

where $U_0 = Ry$ is the hydrogen atom ionization energy and $E_{at} = e/a_B^2$ is the intra-atomic field strength.

Notice that the governing parameter of the Keldysh ionization theory [13] is

$$\gamma = \frac{\omega\sqrt{2mU_0}}{eE_0},$$

where U_0 is the atomic electron binding energy. In the case of a hydrogen atom, we get

$$= \frac{me\omega}{\hbar E_0} = \frac{\hbar \omega}{eE_0 a_B}.$$

Thus, we can see that there is the direct coupling between the parameters μ_0 and γ: $\mu_0 \gamma = 1$.

In previous work [49], we have calculated the dependence of a silver atom ionization probability on the laser pulse field strength and made comparison with the Keldysh theory. The results of comparison have shown that if in computer simulations we shall take into account only the ground state, then the obtained dependence exactly coincides with the one that predicted by Keldysh theory. On the one hand, it is quite expectable because we have compared our results with the Keldysh's rate of ionization from the ground atomic state. On the other hand, it has been shown [49] that in the case of the ultrashort laser pulses, we could not ignore the multilevel structure of the bound atomic states even in the absence of resonances between the laser pulse carrier frequency (or its harmonics) and energy level spacing of atomic states.

In the far-field range, the spectrum of the vector potential of an atomic response field is given by

$$\mathbf{A}_r(\mathbf{r}, \omega) = \frac{\exp(ikr)}{rc} \int j(\mathbf{r}', \omega) \exp(-ik\mathbf{r}')\, dV',$$

where $\boldsymbol{j}(\boldsymbol{r},\omega)$ is the spectrum of an atomic current density, which is defined by the well known expression [60]

$$\boldsymbol{j}(\boldsymbol{r},t)=\frac{q}{2m}\left[\psi^{*}\cdot\left(\boldsymbol{p}-\frac{q}{c}\boldsymbol{A}\right)\psi+\left(\left(\boldsymbol{p}-\frac{q}{c}\boldsymbol{A}\right)\psi\right)^{*}\cdot\psi\right]. \qquad (6.16)$$

In its turn, the spectral density of response field intensity is given by

$$\frac{dI}{d\omega}=\frac{c}{4}\left|B_r(\boldsymbol{r},\omega)^2\right|r^2do=\frac{\omega^2}{4c^3}\left|\int\left[\boldsymbol{j}(\boldsymbol{r}',\omega)\,n\right]\exp\left(-ik\boldsymbol{r}'\right)dV'\right|^2do. \qquad (6.17)$$

In the case when the long-wave approximation, $ka_0 \ll 1$ (where a_0 is the amplitude of an electron oscillation in an external field) holds for all field harmonics, the intensity spectrum of a single-atom response is given by

$$\frac{dI}{d\omega}=\frac{\omega^2}{4c^3}\left|\left[\boldsymbol{J}(\omega)\,n\right]\right|^2do,$$

where

$$\boldsymbol{J}(\omega)=\int\boldsymbol{j}(\boldsymbol{r},\omega)\,dV.$$

However, if we deal with the spatially distributed ensemble of atoms, Eq. (6.17) takes the form

$$\boldsymbol{J}(\boldsymbol{k},\omega)=\sum_{i=1}^{N}\boldsymbol{J}_i(\omega)\exp\left[i\left(\boldsymbol{k}\boldsymbol{r}_i-\omega t_i\right)\right], \qquad (6.18)$$

where r_i are the coordinates of the different atoms of an ensemble. Let $\boldsymbol{k}_0$ be the wave vector of the incident electromagnetic wave. Then, the phase of the field in the position of i-th atom is $\omega_0 t-\boldsymbol{k}_0\boldsymbol{r}_i=\omega_0(t-t_i)$, where $t_i=\boldsymbol{k}_0\boldsymbol{r}_i/\omega_0$. By substituting t_i into Eq. (6.18), we get

$$\boldsymbol{J}(\boldsymbol{k},\omega)=\sum_{i=1}^{N}\boldsymbol{J}_i(\omega)\exp\left[i\frac{\omega}{c}\left(\boldsymbol{n}(\omega)-\boldsymbol{n}_0(\omega_0)\right)\boldsymbol{r}_i\right], \qquad (6.19)$$

where $\boldsymbol{n}(\omega)=\boldsymbol{k}c/\omega$ and $n_0(\omega_0)=\boldsymbol{k}_0c/\omega_0$. It is seen that in the case of the spatially distributed ensemble of atoms, the dispersive properties of medium start playing the important role.

The matrix elements of the generalized momentum operator $\boldsymbol{P}=\boldsymbol{p}-q\boldsymbol{A}/c$ in the bases of the eigenfunctions (6.3) and (6.10) appear as

$$\int u_n^*(\boldsymbol{r})\left(p-\frac{q}{c}\boldsymbol{A}\right)u_m(\boldsymbol{r})\,dV=\sum_{k,p}V_{nk}^{-1}(t)\,\boldsymbol{p}_{kp}\,V_{pm}(t), \qquad (6.20)$$

$$\int \varphi_n^* (\mathbf{r}, t) \left(\mathbf{p} - \frac{q}{c} \mathbf{A} \right) \varphi_m (\mathbf{r}, t) \, dV = \int u_n^* (r) \, \mathbf{p} u_m (r) \, dV = p_{nm} . \tag{6.21}$$

By substituting the expansion (6.11) into (6.16) and applying Eq. (6.17), we get the following equation for atomic current

$$\mathbf{J}(t) = \frac{q}{m} \sum_{n,m,p,q} a_n^* (t) \, a_m (t) \, V_{np}^{-1} (t) \, p_{pq} \, V_{qm} (t) . \tag{6.22}$$

The most principle feature of Eqs. (6.20) and (6.21) is that the right-hand side of these equations includes the matrix elements $\mathbf{p}_{nm}$, which are calculated on the basis of the "free atom" eigenstates. Notice that on the basis of the "free atom" eigenstates, the matrix elements of momentum operator are strongly related with the coordinate operator $\mathbf{p}_{nm} = im\omega_{nm} \mathbf{r}_{nm}$ (see, e.g., [61]).

This relation demonstrates the further advantages of the basis of the eigenfunctions (6.10) for calculation of matrix elements of quantum mechanical operators.

Using the obtained relation, we can rewrite (6.22) in the following identical form

$$\mathbf{J}(t) = i \sum_{n,m,p,q} a_n^* (t) \, a_m (t) \, \omega_{pq} V_{np}^{-1} (t) \, d_{pq} \, V_{qm} (t) . \tag{6.23}$$

So, for calculation of the spectrum of the atomic response field, we should solve numerically the set of equations (6.12) for probability amplitudes a_n.

6.1.4 *Numerical Research*

In numerical simulations, we assume that a single silver atom interacts with the pulse of Ti:sapphire laser (the wavelength $\lambda = 800$ nm). The time profile of the pulse is defined by Eq. (6.14), and the pulse duration is assumed as $\tau = 27$ fs. The amplitude of the laser field strength E_0 is varied in the wide range starting from a significantly subatomic to near-atomic values. As we have discussed in our previous work [49], if the laser field strength lies in the interval, $0 \leq \mu_0 \leq 2.5$ then the approximately complete basis of states forms the following 284 levels of a silver atom: 5s (GS), 5p, 5d, 6p of the discrete spectrum and 280 states of continuum spectrum with orbital quantum numbers $l = 0 - 3$. In order to optimize the

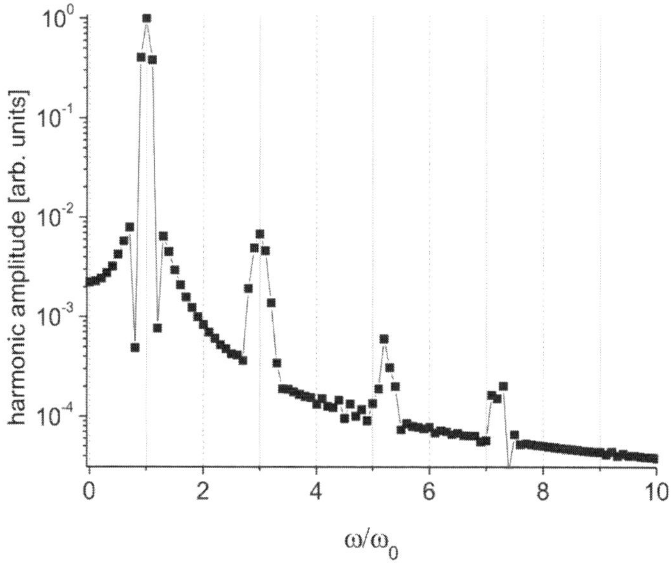

Figure 6.5 The photoemission spectrum generated in silver at laser pulse intensity $I = 5.94 \times 10^{10}$ W/cm^2. Reproduced from Ref. 1 with kind permission from The European Physical Journal.

range and the step in energy space of continuum band, we have previously calculated the matrix elements V_{nm} for above-mentioned discrete-continuum transitions (see for details [49]).

Our numerical calculations yield a series of harmonic emission spectra. In subatomic region, the spectrum consists of the sequence of the odd harmonics of incident field carrier frequency (Fig. 6.5). The amplitudes of harmonics sharply drop with the harmonic number. Such shape of spectra is quite predictable from Eq. (6.22). Indeed, the matrix elements V_{nm}, as it follows from Eq. (6.8), represent the infinite power series of the field parameter μ_0. This series consists of even powers of μ_0 (i.e., μ_0^{2n}), when the states n and m have the identical parity, and odd powers, μ_0^{2n+1}, when the states n and m have the opposite parity. In the substantially subatomic region of a field strength, the ground state is maximally populated, hence, $|a_0|^2 \approx 1$ and $\left|a_{n\neq 0}\right|^2 \approx 0$. So, in this case, Eq. (6.23) gets the form

$$\boldsymbol{J}(t) = i \sum_{p,q} \omega_{pq} V_{0p}^{-1}(t)\, \boldsymbol{d}_{pq}\, V_{q0}(t)\,.$$

The matrix elements $\boldsymbol{r}_{pq}$ have non-zero values only for transitions between states with the opposite parity, so one of the matrix elements V_{0p}^{-1} and V_{q0} include only even harmonics and another odd. Hence, in the region of subatomic field strength, the response spectrum includes the odd harmonics only.

The above discussion demonstrates the principle difference between the properties of the eigenstates of the bases (6.2) and (6.4). The eigenfunctions of the free atom boundary value problem do not depend on time. Hence, the averages of the quantum mechanical operators can be time dependent only in the case when the population of the different atomic states varies with time. An atom could not emit radiation for the whole time in the ground state. It means that if we use the basis of free atom eigenfunctions, then the response field can appear only as a result of inelastic processes of the light–atom interactions when atom comes to the excited states by absorbing the energy of electromagnetic field. On the contrary, the eigenfunctions of the boundary value (6.4) depend on time, whereas the eigenvalues remain invariable. As a result, an atom, being the whole time in the ground state of energy spectrum of the boundary value problem (6.4), can emit a radiation.

The origin of this emission is the light-induced (and, hence, time-dependent) anisotropy of a spatial profile of an electron wave function: the electromagnetic wave induces the radial electron motion in the intra-atomic potential field and, hence, produces some external work. This motion will inevitably result in the appearance of harmonics in the response field, if the intra-atomic potential is not harmonic one. At a given laser field strength, this process could be the process of an elastic light–atom interaction, because after the termination of the laser pulse action, the atomic electron can be again found in the ground state.

When the laser field strength approaches $\mu_0 \approx 10^{-2}$ ($I \approx 3 \times 10^{11}$ W/cm^2), the shape of the response spectrum is changed. There appear tendencies for growing of its width and formation of plateau (Fig. 6.6). Further increase in the amplitude of the external field strength allows us to observe well-defined plateau with definite CF [see Fig. 6.7 for $\mu_0 = 0.08$($I = 1.88 \times 10^{13}$ W/cm^2)].

To give some interpretation of such modification of the response spectra, let us turn to some analytical evaluations of the solutions of

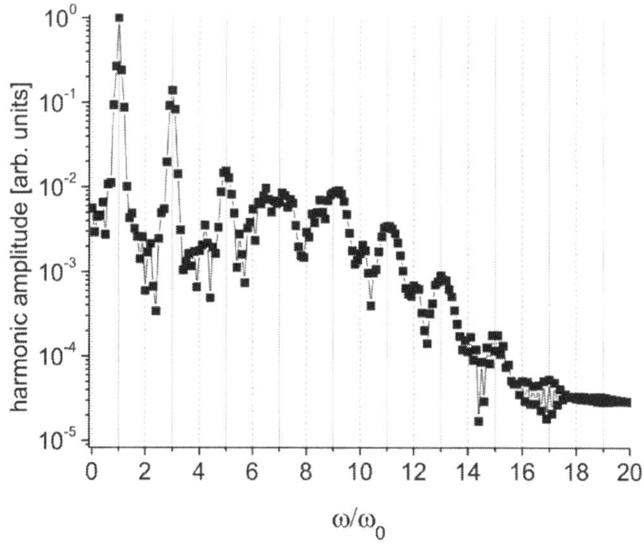

Figure 6.6 The photoemission spectrum generated in silver at laser pulse intensity $I = 1.83 \times 10^{12}$ W/cm^2. Reproduced from Ref. 1 with kind permission from The European Physical Journal.

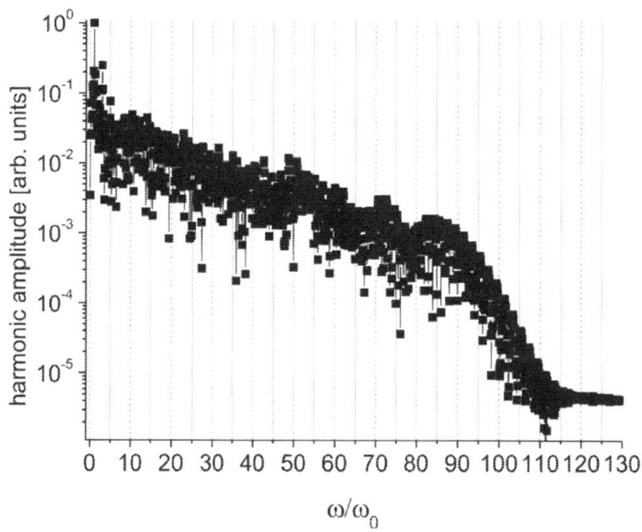

Figure 6.7 The photoemission spectrum generated in silver at laser pulse intensity $I = 1.88 \times 10^{13}$ W/cm^2. Reproduced from Ref. 1 with kind permission from The European Physical Journal.

Eqs. (6.12). In the absence of resonances between the fundamental field frequency (as well as its harmonics) and the frequencies of atomic transitions, the population of the atomic states can be calculated in frames of perturbation theory. Attributing the indexes n_0 and n_1 to the ground and some excited states, respectively, in the frame of the first-order perturbation theory approximation, we get the following solutions for the probability amplitudes

$$a_{n_0}(t) = a_{n_0} exp\left[-i\Phi_0(t)\right],$$
$$a_{n_1}(t) = a_{n_0} exp\left[-i\Phi_1(t)\right] \int_{-\infty}^{t} M_{n_1 n_0}(t') \, exp\left[i\left(\Phi_1(t') - \Phi_0(t')\right)\right] dt', \tag{6.24}$$

where

$$\Phi_i(t) = \frac{1}{\hbar} \int_{-\infty}^{t} M_{n_i n_i}(t') \, dt'$$

and M_{nm} are the compound matrix elements:

$$M_{nm} = \sum_k V_{nk}^{-1} E_k V_{km}.$$

The angular part of the matrix elements V_{nm} is calculated in an explicit analytical form for any atom with spherically symmetric intra-atomic potential. In this case, the radial wave functions depend on the principal quantum number n and angular momentum l, and the radial part of matrix elements V_{nm} takes the following form

$$\left\langle n_2 l_2 \left\| j_l\left(\frac{q}{\hbar c} A(t) r\right) \right\| n_1 l_1 \right\rangle = \int_0^{\infty} R_{n_2 l_2}(r) \, j_l\left(\frac{q}{\hbar c} A(t) r\right) R_{n_1 l_1}(r) \, r^2 dr.$$

where $j_l(z)$ are the spherical Bessel functions. Notice that in accordance with the angular momentum summation rules, the indexes in the last equation obey the following condition: $|l_2 - l_1| \leq l \leq l_1 + l_2$. As we have mentioned above, for any hydrogen-like atom, these matrix elements can be calculated in an explicit analytic form. It is easily seen from Eqs. (6.12) that the diagonal compound matrix elements $M_{nn}(t)$ determine the instant energy of an electron, which is in the free atom eigenstate $u_n(\vec{r})$. It should be reminded that if we describe the atomic electron evolution using the basis of the free atom eigenfunctions, then the energies of the states depend on the field amplitude, and, as result, they vary in time.

To avoid the overcomplicated mathematical expressions for definition of the CF, let us concentrate on the hydrogen atom. In the subatomic region of the laser field strength, we restrict ourselves by account of the two lowest discrete states, for which we have

$$\langle 1s \,\|\, j_0 \,\|\, 1s\rangle = \frac{16}{\left(4+\mu^2(t)\right)^2}, \quad \langle 2p \,\|\, j_0 \,\|\, 2p\rangle = \frac{1-\mu^2(t)}{\left(1+\mu^2(t)\right)^4}. \tag{6.25}$$

In the subatomic region, $\mu_0 \ll 1$, we can use the quadratic approximation of the matrix elements (25), then for the compound matrix elements M_{nn} we get

$$M_{2p}(t) \approx E_{2p}\left(1-10\mu^2(t)\right), \quad M_{1s}(t) \approx E_{1s}\left(1-\mu^2(t)\right). \tag{6.26}$$

Here, we have signed $M_n = M_{nn}$ for brevity. Assuming that incident pulse profile has the form

$$A(t) = A_0 \exp\left(-\frac{t^2}{\tau_0^2}\right)\cos(\omega_0 t)$$

and executing time integration, we find the expression for $\Phi_n(t)$

$$\Phi_{2p}(t) - \omega_{2p}t =$$

$$= -\omega_{2p}\sqrt{\frac{\pi}{2}}\frac{5\mu_0^2\tau_0}{4}\left[2\mathrm{erf}\left(\frac{\sqrt{2}t}{\tau_0}\right) + i\exp\left(-\frac{\omega_0^2\tau_0^2}{2}\right)\right.$$

$$\left.\times\left(\mathrm{erfi}\left(\frac{-2it+\omega_0\tau_0^2}{\tau_0}\right) - \mathrm{erfi}\left(\frac{2it+\omega_0\tau_0^2}{\tau_0}\right)\right)\right] \tag{6.27}$$

$$\Phi_{1s}(t) - \omega_{1s}t =$$

$$= -\omega_{1s}\sqrt{\frac{\pi}{2}}\frac{\mu_0^2\tau_0}{8}\left[2\mathrm{erf}\left(\frac{\sqrt{2}t}{\tau_0}\right) + i\exp\left(-\frac{\omega_0^2\tau_0^2}{2}\right)\right.$$

$$\left.\times\left(\mathrm{erfi}\left(\frac{-2it+\omega_0\tau_0^2}{\tau_0}\right) - \mathrm{erfi}\left(\frac{2it+\omega_0\tau_0^2}{\tau_0}\right)\right)\right], \tag{6.28}$$

where $\mathrm{erf}(z)$ is the error function and $\mathrm{erfi}(z) = \mathrm{erf}(iz)/i$. The first term in square brackets describes energy shift depending on pulse envelope and the second one shows oscillations with a double carrier frequency. The second term equals exactly zero for the external pulse of rectangular profile, and it is negligibly small for the case of Gaussian pulse profile if its duration satisfies the condition $\omega_0 \tau_0 \gg 1$. Hence, in subatomic region ($\mu_0 \ll 1$) the phases $\Phi_i(t)$ can be approximated as

$$\Phi_{2p}(t) = \omega_{2p}\left[t - \frac{5}{2}\sqrt{\frac{\pi}{2}}\mu_0^2\tau_0 \tanh\left(\frac{\sqrt{3}t}{\tau_0}\right)\right],$$

$$\Phi_{1s}(t) = \omega_{1s}\left[t - \frac{1}{4}\sqrt{\frac{\pi}{2}}\mu_0^2\tau_0 \tanh\left(\frac{\sqrt{3}t}{\tau_0}\right)\right].$$

Therefore, during the laser pulse action, the energy-level spacing depends on time and the instantaneous frequency reads as

$$\begin{aligned}\Omega(t) &= \frac{d}{dt}\left(\Phi_{2p}(t) - \Phi_{1s}(t)\right)\\ &= \omega_{2p} - \omega_{1s} - \sqrt{\frac{3\pi}{2}}\mu_0^2\left(\frac{5}{2}\omega_{2p} - \frac{1}{4}\omega_{1s}\right)\cosh^{-2}\left(\frac{\sqrt{3}t}{\tau_0}\right).\end{aligned}$$

It is seen that the magnitude of energy shift is

$$\Delta\omega = \sqrt{\frac{3\pi}{2}}\mu_0^2\left(\frac{5}{2}\omega_{2p} - \frac{1}{4}\omega_{1s}\right). \tag{6.29}$$

Energy shift $\Delta E = \hbar|\Delta\omega|$ in a hydrogen-like atom for transition $1s \to 2p$ is found then to be

$$\Delta E = \sqrt{\frac{3\pi}{2}}\frac{3}{4}U_\mathrm{p},$$

where U_p is a ponderomotive potential, given by the well-known expression

$$U_\mathrm{p} = \frac{e^2E_0^2}{4m\omega^2}.$$

By substituting Eq. (6.24) into Eq. (6.23), we can see the difference between the diagonal [$n = m$ in Eq. (6.23)] and non-diagonal ($n \neq m$) partial matrix elements of the atomic current. We have seen in previous subsection that the diagonal elements result in a sequence of odd harmonics. At the same time, the shape of spectra prescribed by the non-diagonal elements depends significantly on the laser pulse spectral width. In the case of a few-cycle laser pulses, this part of the integral spectrum takes the shape of the quasi-continuum plateau, spread till double energy shift of Eq. (6.29), instead of a sequence of distinct harmonics. Nevertheless, in the subatomic region of a field strength, the cutoff energy, E_c, can be estimated as

$$E_c = \frac{3}{2}\sqrt{\frac{3\pi}{2}}U_p = 3.26 \cdot U_p.$$

It is seen that the obtained equation coincides approximately with the well known equation $E_c = U_0 + 3.17 \cdot U_p$ proposed in [22, 62] and calculated in [47]. The same calculations can be easily made for a silver atom.

Equation (6.29) is true only in subatomic region, that is, at $\mu_0 \ll 1$, when the approximation (6.26) for matrix elements (6.25) is valid. As it has been shown in our previous works [49, 51–53], the non-diagonal reduced matrix elements reach maximum at $0.1 \leq \mu_0 \leq 1$, whereas the diagonal matrix elements drop sharply in this region. This is the main reason why at $\mu_0 \geq 1$ the incident field strengthening leads to saturation of the CF. Such behavior of the CF is confirmed by both numerical calculations (see Fig. 6.8) and experimental measurements.

The results of computer calculations of the CF as a function of the laser pulse intensity are presented in Fig. 6.8. The solid line in this figure shows the approximated analytical dependence obtained for a silver atom in the frames of μ^2 approximation [see Eqs. (6.25)–(6.29)]. In the region of the laser pulse intensity of $\left(10^{11} \leq I \leq 10^{12}\right)$ W/cm^2, the results of computer calculations are in good agreement with the approximate equation. This agreement is due to the following: the solid line (see Fig. 6.8) has been derived under account of only two discrete states, namely 5s and 5p. On the contrary, in this region of pulse intensity, the population of 5p state exceeds significantly the population of other excited states of discrete spectrum, as it has been shown in our previous works [49, 54], in which the photoelectron spectra of a silver atom have been studied. Under further increase in the laser pulse intensity, the calculated curve raises faster than it is predicted by the equation obtained in μ^2 approximation. In this region of pulse intensity, the population of other discrete spectrum excited states becomes comparable and even exceeds the population of 5p state due to the

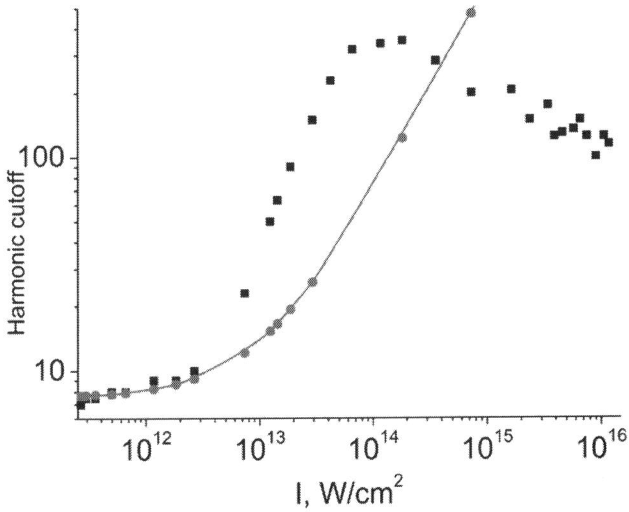

Figure 6.8 Cutoff frequency as a function of the laser pulse intensity. Squares–is a result of the numerical calculations, solid line with circles–dependence given by Eqs. (6.25–6.29). Reproduced from Ref. 1 with kind permission from The European Physical Journal.

recombination process. Hence, the new channels of photoemission are revealed.

At laser pulse intensity $I > 10^{14}$ W/cm^2, the CF is saturated, that is, it ceases to be intensity dependent. The reason of such behavior is quite obvious and it is associated with the fact that the probability of an electron ionization approaches to unity in this region of a pulse intensity. It means that the atomic electron is almost entirely localized in the continuum spectrum states and it does not collide with its parent ion. At the same time, as far as the time profile (6.14) of a laser pulse has relatively soft slope in the front of pulse, hence, the harmonics are effectively generated here. So, the results of computer modeling show that the most probable reason of the CF saturation is the total ionization of the irradiated atom.

6.1.5 *Discussion*

One can see from Fig. 6.8 that the calculated laser pulse intensity, $I_s^{(t)} \approx 10^{14}$ W/cm^2, at which the CF is saturated, is close to the experimentally measured value, $I_s^{(e)} = (4 \pm 1) \times 10^{14}$ W/cm^2. However, the calculated maximal order of harmonics exceeds the experimentally measured value for more than four times. In computer simulations, we deal with a single silver atom. Hence, it is natural to assume that the reason of this difference can be due to the frequency-dependent dispersion appropriate to the spatially distributed ensembles of atoms (see, e.g., [63, 64]). To illustrate qualitatively the difference between the frequency-angular spectra of a single atom response and an ensemble of atoms, we shall use further the simplest dispersion model.

The frequency–angular response spectrum for spatially distributed ensemble of atoms is given by Eq. (6.20). According to Eq. (6.22) and taking into account a decomposition of $V = \exp\left(i \frac{q}{\hbar c} \vec{A}\vec{r}\right)$ into a series of the laser field powers, the field of an atomic response at the h-th harmonic ($\omega_h = h\omega_0$) frequency is determined by an infinite series of the laser field powers, such as as $A(\omega_h) \sim \sum_m C_h^{(m)} A_0^{h \pm m} A_0^{m}$ (see for details [49, 53–55]). However, in the subatomic field strength, the main contribution is given by the term $A_0^h(\boldsymbol{r}, t)$. So, if the atoms of an ensemble are identical, then Eq. (20) can be rewritten in the following form

$$\boldsymbol{J}(\boldsymbol{k}, \omega) = \sum_{i=1}^{N} \boldsymbol{J}_i(\omega) \exp\left[i \frac{\omega}{c} \left(\boldsymbol{n}(\omega) - \boldsymbol{n}_0(\omega_0)\right) \boldsymbol{r}_i\right] = \boldsymbol{J}_0(\omega) f(\boldsymbol{k}, \boldsymbol{k}_0),$$

where $\boldsymbol{J}_0(\omega)$ is the spectrum of a single atom current and the form factor $f(\boldsymbol{k}, \boldsymbol{k}_0)$ is defined by

$$f(\boldsymbol{k}, \boldsymbol{k}_0) = \int \rho_\omega(\boldsymbol{r}) \exp\left[i \frac{\omega}{c} \left(\boldsymbol{n}(\omega) - \boldsymbol{n}_0(\omega_0)\right) \boldsymbol{r}\right] dV. \tag{6.30}$$

Here, $\rho_\omega(\vec{r})$ is an effective density of responding atoms, which is the product of density of atoms and the h-th power of the amplitude spatial distribution of driving laser pulse in irradiated volume.

According to the Lorentz classical theory, the refractive index for electromagnetic wave propagating in ionized gaseous medium is given by

$$n(\omega) \approx 1 - \frac{1}{2}\left[\sum_\alpha \frac{\omega_{p\alpha}^2}{\omega^2 - \omega_{0\alpha}^2 + i\omega\gamma_\alpha} + \frac{\omega_{pe}^2}{\omega^2 + i\omega\gamma_e}\right], \tag{6.31}$$

where

$$\omega_{p\alpha(e)} = \sqrt{\frac{4\pi e^2 N_{\alpha(e)}}{mV}}$$

is the plasma frequency corresponding to neutral atoms ($\alpha = I$), ions ($\alpha = II$), and free electrons (e); N_α is the number of neutral atoms and ions in the irradiated volume V, and N_e is the number of free electrons. When neutral and ionized atoms stay mainly in the ground state, we can assume that $\hbar\omega_{0I(II)}$ are the binding energies of a silver atom ($\alpha = I$) and ion ($\alpha = II$), respectively. In the case of ultrashort laser pulses, the relaxation rates $\gamma_{\alpha(e)}$ are inversely proportional to the laser pulse temporal width.

Thus, Eq. (6.31) takes into account the dispersion due to the atomic neutral or ionized gas and free electrons. If the medium is the neutral gas in the initial moment, then we have $N_I = N_0$, $N_{II} = 0$, and $N_e = 0$. In the plasma state, appearing in the process of the light-atom interaction, the density of the different species is varied in time and we get $N_I = N_0\,(1 - p\,(t))$, $N_{II} = N_0\,p\,(t)$, and $N_e = N_0\,p\,(t)$, where $p\,(t)$ is the probability of ionization. So, the refractive index is really a slowly varying function of time; the characteristic time of variation is determined by the laser pulse temporal width.

In the case of the Gaussian incident beam, the effective density of responding atoms can be approximated as follows

$$\rho_\omega\,(\boldsymbol{r}) = \frac{N_0}{V}\left[\exp\left(-\frac{z^2}{L^2} - \frac{\rho^2}{d^2}\right)\right]^{\omega/\omega_0}, \qquad (6.32)$$

where d is the focal waist of a driving laser beam and L depends both on the confocal parameter of driving laser beam and a spatial distribution of an atomic density in the laser plume produced by the pump laser pulse. We have assumed that the laser beam is propagated along the z axis.

The presented model enables us to estimate, at least qualitatively, how the dispersion effects will influence on the profile of a response spectrum. It is well known that in the case of a neutral gas medium, the refraction index varies significantly in the visible and near UV regions and approaches to unity in the XUV and X-ray regions. As it follows from Eq. (6.31), in the subatomic region of a laser field strength, when $p\,(t) \ll 1$, the influence of the dispersion on the emission spectrum of an atomic ensemble is mainly associated with the neutral atoms, because the susceptibility is positive for a driving laser pulse and negative for high order harmonics. On the contrary, in the case of highly ionized media, when the probability $p(t)$ approaches to unity, the phase mismatch is completely due to the plasma dispersion. Indeed, in this limit, the refractive index became

$$n(\omega) \approx 1 - \frac{\omega_p^2}{2\omega^2}\left(1 + \frac{1}{1 - \omega_{0II}^2/\omega^2}\right).$$

The susceptibility is negative for both driving laser pulse and high-order harmonics. The coherence length L_c can be estimated from the following condition

$$\frac{\omega}{c} L_c \Delta n = \pi,$$

where $\Delta n = n(\omega) - n(\omega_0)$; at near atomic field strength, we have $\Delta n \approx \omega_p^2/2\omega_0^2$. Hence, for the coherent length, we get

$$L_c = \frac{4\pi^2}{h\lambda_0}\left(\frac{\omega_p}{c}\right)^{-2}.$$

So, the coherent length is inversely proportional to the harmonic order h. However, it should be noted that due to inhomogeneous distribution of driving laser pulse intensity in irradiated area, the effective length of interaction is different for different harmonics, and according to Eq. (6.32), we can estimate it as follows: $L_\omega = L/\sqrt{h}$. As a result, the coherence length falls with harmonic order more slowly, $L_c(\omega) \propto 1/\sqrt{h}$.

Thus, the dispersion properties of a spatially distributed ensemble of atoms could result in drastic transformation of atomic ensemble emission spectra in comparison with the single atom one. By taking into account the experimental conditions described in subsection 6.1.2, we have calculated the function $\left|f\left(\vec{k}, \vec{k}_0\right)\right|^2$ defined by Eq. (6.30). Figure 6.9 shows in comparison the single atom response spectrum (Fig. 6.9a) and spatially distributed ensemble of atoms (Fig. 6.9b). Figure 6.9 corresponds to the near-atomic field strength, therefore to take into account the temporal dependency of the ionization probability, we have made averaging over $0 \le p \le 1$ under calculation of $f\left(\vec{k}, \vec{k}_0\right)$. One can see that the CF in atomic ensemble spectrum is significantly shifted in the long wave region in comparison with the CF in spectrum of single

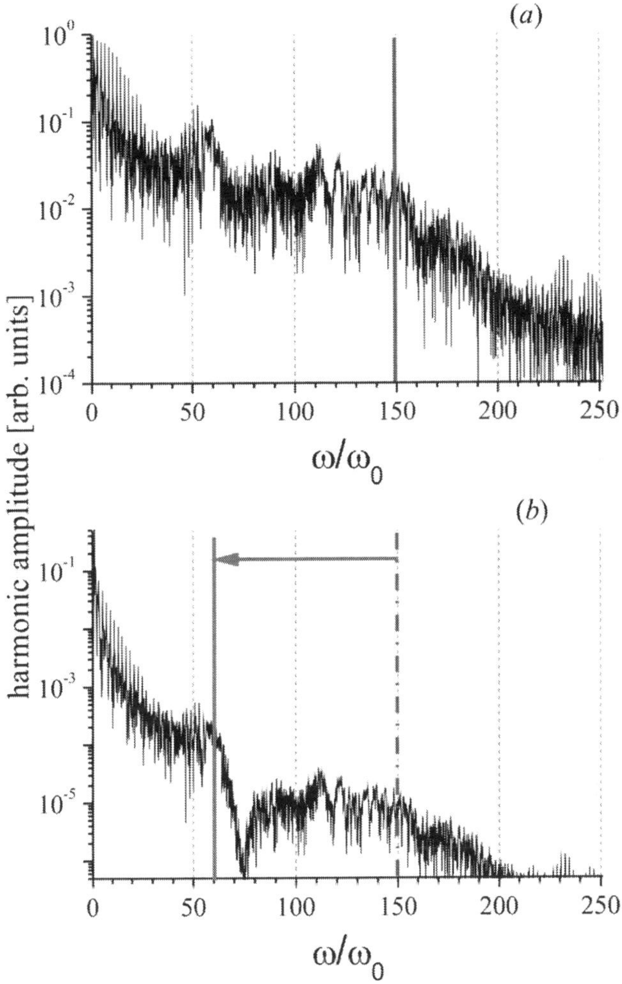

Figure 6.9 The comparison of photoemission spectra generated in silver by a single atom (a) and an ensemble of atoms (b) at the laser pulse intensity $I = 2.37 \times 10^{15}$ W/cm^2. Reproduced from Ref. 1 with kind permission from The European Physical Journal.

atom response. Figure 6.10 shows the integral dependency of the CF as a function of laser pulse intensity, which is calculated for spatially distributed ensemble of atoms. In these calculations, we have taken into account that the ionization probability depends on

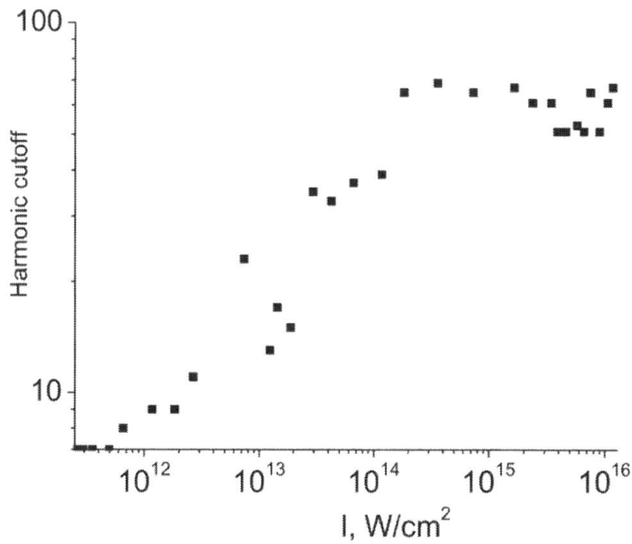

Figure 6.10 Cutoff frequency of an ensemble of atoms as a function of a laser pulse intensity. Reproduced from Ref. 1 with kind permission from The European Physical Journal.

the peak pulse intensity (see [49] for details). It can be seen that the curve in Fig. 6.10 fits much closer to the experimentally measured dependency.

In the case of a divergent laser beam, in accordance with Eqs. (6.30) and (6.32), the partial angular components of form factor can be approximated as follows

$$f_{\omega}(\theta, \theta_0) = \exp\left\{-\frac{1}{4}\left[kL_{\omega}(n\cos\theta - n_0\cos\theta_0)\right]^2 - \frac{1}{4}\left[kd_{\omega}(n\sin\theta - n_0\sin\theta_0)\right]^2\right\}, \qquad (6.33)$$

where $L_{\omega} = L/\sqrt{h}$ and $d_{\omega} = d/\sqrt{h}$. It is seen that in the case dispersion free medium (i.e., $|\vec{n}(\omega)| = |\vec{n}_0(\omega_0)| = 1$), the peak intensity of harmonic emission is always on the axis of laser beam. In the case of a dispersive medium, the direction of harmonic peak intensity is determined by the following condition: $n(\omega)\cos\theta_m^{(h)} = n_0\cos\theta_0$. In the case of the plain laser wave (i.e., $\theta_0 = 0$), we get the conical emission for harmonics. The intensity of different harmonics reaches maximum at different angles with respect to the direction of the laser pulse wave-vector.

Due to temporal dynamics of ionization probability, we can really speak only on the broadening or narrowing of the angular width of harmonic emission. Indeed, in the case of the subatomic field strength, when the ionization probability is small, the susceptibility for a low-order harmonics is positive, that is, $n_0(\omega_0) \geq 1$; hence, the angular width of harmonic emission is narrowed in comparison with the laser beam width, because $\theta_m^{(h)} < \theta_0$. On the contrary, in the intense laser field, the susceptibility becomes negative, hence $n_0(\omega_0) \leq 1$. In this case, for an apex angle of the high-order harmonic conical emission, we get

$$\theta_m^{(h)} \approx \sqrt{\theta_0^2 + 2\frac{\omega_p^2}{\omega_0^2}\left(1 - \frac{1}{h^2}\right)}.$$

So, in the intense laser field, the angular width of the harmonic emission is broadened in comparison with the laser beam width. Notice that in more precise analysis of angular spectra, we should also take into account the second term in the exponent of (6.33), because it depends strongly on the angle mismatch.

The above discussion provides at least the qualitative interpretation of results presented in Fig. 6.2a,b. The decrease in intensity of driving laser pulse due to aperturing of incident beam results in the increase of dispersion associated with the neutral atoms. Indeed, the low-order harmonics are mainly emitted in the front of the driving laser pulse, when the probability of ionization is small. We have mentioned above in this case the angular width of the harmonic emission is narrowed. At the same time, the angular width of emission of the high-order harmonics is broadened, because they are primarily emitted in the body of driving laser pulse, when the medium is highly ionized.

6.2 Calculations of Plasma Formation for Harmonics Generation

Concentration of particles in gases and laser-ablated plasmas is an important factor, which greatly influences the efficiency of HHG, because a single atom during strong-field HHG can emit only one photon of harmonic radiation. The increase of particle concentration in general increases the efficiency of pump radiation absorption, which means that the enhanced number of electron recombination acts and, as a result, results in the enhanced HHG yield. However, the increase of particle concentration is not always useful, because, at high particle concentration, when the radiation is completely absorbed, overall pulse energy may be insufficient to generate even a single harmonic of maximum order given by the cutoff law $E_{\max} = I_p + 3.17U_p$ [41, 47, 62]. This means that if $E_{\text{pulse}}/N < E_{\max}$, HHG efficiency in cutoff domain can be reduced (here $E_{\max}$ is the maximum energy of photon at cut-off region, I_{p} is the ionization potential of emitter, U_{p}. is the ponderomotive energy of accelerated electron, E_{pulse}/N is the pulse energy, and N is the total number of irradiated particles). Actually, the destructive influence of concentration on HHG yield in the cutoff region can begin even earlier, because pump pulse is less intense at the beginning and in the end, so the last relation is a good estimate quantity. One should note that N is a number of particles, not single atoms, so in the case of nanoparticle formation, this estimate can be changed. Thus, the determination of particle concentration in the interaction region is an important step for exact quantitative estimates of HHG. However, while for gas jets the determination of concentration during experiment is still possible, although is also rather difficult taking into account a complicated structure of modulated gas jets, the experimental determination of laser-ablated plasma concentration in experiment is a difficult task.

There have been various efforts to create more efficient HHG sources, including techniques using gas and plasma media, as well as cluster media, which enhance harmonic intensity [66–68], modulated hollow-core waveguides for adjusting phase matching [69], quasi-phase matching with several gas jets [70], and resonance-induced enhancement [71]. These investigations led to generation of microJoule level harmonic pulses both in gases [72] and carbon plasmas [73, 74]. The promising earlier results of HHG from the carbon-containing laser plasma plumes comprising fullerenes [75], carbon nanotubes [76], and graphite [73, 74] in the 40–100 nm range, where considerable enhancements in harmonic yield were observed, prompt further investigations. These observations have motivated us to compare the HHG efficiencies in carbon and metal-containing plasmas and gases under closely matched conditions. Utilization of similar optimal conditions allowed investigating the influence of concentration on the HHG with good certainty.

We use the molecular dynamics (MD) simulation below to calculate the concentration dependence on various pulse parameters. In this section, we analyze the relations between concentration dependence and HHG behavior in carbon and other plasmas, while also studying the process of nanoparticle formation in more detail.

6.2.1 *Theory*

Although it is possible to predict material properties from ab initio simulations to high accuracy, the method has an immense drawback. The large computational times required for these calculations make them unpractical for typical simulations of laser ablation. The system sizes, which can be simulated on modern computers, are still in the order of a few hundreds of atoms, which is way too small. In the case of molecular dynamics simulations, the system is simplified by not treating the ions and electrons separately. The modeled atoms do not have inner degrees of freedom anymore; they interact as classical particles with each other. For the solution of the many-body problem, the classical Hamiltonian equations of motion are integrated. In general, for an N-particle system, this leads to $6N$ first-order differential equations. Let

$$H(p_1, \ldots, p_N, r_1, \ldots, r_N) = \sum_{i=1}^{N} \frac{p_i^2}{2m_i} + U(r_1, \ldots, r_N) \tag{6.34}$$

be the Hamiltonian of a configuration of N classical particles, $U(\{r_i\})$ being the potential. The equations of motion then become:

$$\dot{r}_j = \frac{p_j}{m_j}, \; \dot{p}_j = F_j = -\nabla_{r_j} U(\{r_i\}) \tag{6.35}$$

The force $F_j = \dot{p}_j$, acting on atom j, is calculated by taking the gradient of the potential U with respect to the coordinate r_j. These equations cannot be solved analytically for large N. Instead, the equations are discretized in time: for a given configuration (coordinates and momenta) at time t, the forces on all the atoms are calculated. After this, their positions and momenta are adjusted accordingly for $t + \delta t$.

The time step δt is not arbitrary. It has to be suitable for the underlying problem. Femtoseconds or even less are necessary for atomistic simulations. The time step has to be smaller than the time scales of characteristic motions to be resolved in the system. This dictates an upper bound for δt. For atomistic simulations of solids, a time step between 0.1 and 2 fs has been established. In this work, the Verlet algorithm is used.

Solving these standard Hamilton equations characterizes a system, which does not exchange particles (N), volume (V) or energy (E) with its environment. In statistical mechanics, a system like this is called microcanonical ensemble (NVE ensemble). It should be noted that, during laser ablation, total energy of the system is changed, but during single step of integration of the equations of motion, the energy is unchanged, so NVE ensemble is quite sufficient for simulation of laser ablation.

For the solution of Eq. (6.35), a potential has to be specified. A general potential is given by taking multi-body contributions into account. The potential energy can be expanded in a part from an external potential, a pair term, a three-body term, and so on. This leads to

$$U\left(\{r_i\}\right) = \sum_i \phi_1(r_i) + \frac{1}{2}\sum_{\substack{i,j\\ i\neq j}} \phi_2(r_i, r_j) + \frac{1}{6}\sum_{\substack{i,j,k\\ i\neq j\neq k\\ j\neq k}} \phi_3(r_i, r_j, r_k) + \ldots \tag{6.36}$$

For the simulation of carbon atoms in graphite crystals, only empirical interatomic potentials [77] were used:

$$V_{ij} = f_C(r_{ij})[f_R(r_{ij}) + b_{ij} f_A(r_{ij})]$$

$$f_C(r) = \left\{ \begin{array}{l} 1 : r < R - D \\ \dfrac{1}{2} - \dfrac{1}{2}\sin\left(\dfrac{\pi(r-R)}{2D}\right) : R - D < r < R + D \\ 0 : r < R + D \end{array} \right\} \tag{6.37}$$

$$f_{\mathrm{R}}(r) = A\exp(-\lambda_1 r)$$

$$f_{\mathrm{A}}(r) = -B\exp(-\lambda_2 r)$$

$$b_{ij} = (1 + \beta^n \zeta_{ij}^n)^{\frac{1}{2n}}$$

$$\zeta_{ij} = \sum_{k \neq i, j} f_C(r_{ik}) g(\theta_{ijk})$$

$$g(\theta) = \gamma_{ijk} \left(1 + \frac{c^2}{d^2} - \frac{c^2}{[d^2 + (\cos\theta - \cos\theta_0)^2]}\right)$$

The parameters for carbon atoms were chosen from [78] (A = 1393.6 eV, B = 346.74 eV, λ_1 = 3.4879 Å, λ_2 = 2.2119 Å, β = 1.5724 $\times 10^{-7}$, n = 072751, c = 38049, d = 4.3484, R = 1.95 Å, D = 0.15 Å). It should be noted that very short range of potential (2.1 Å) does not allow studying interlayer forces, so the actual ablation yield will be slightly increased. However, the computational effort is greatly reduced, because the number of the bonds is reduced as well and the potential is not a two-body one, because b_{ij} is not a constant. In fact, it is the bond order for the bond joining i and j, and it is a decreasing function of a "coordination" ζ_{ij} assigned to the bond.

If the physical mechanisms involved in the ablation process are not known, a modeling of the macroscopic behavior can be a workaround. For this approach, the temperature near the surface has to be increased during the simulation. The temperature of a system is related to the average kinetic energy of its atoms via

$$\langle E_{\text{kin}} \rangle = \frac{f}{2} k_B T \tag{6.38}$$

where f is the number of degrees of freedom, $f = 3$ in the existing case. The basic idea of the direct approach is a rescaling of the kinetic energy of the atoms. After each time step δt, the momentum vector p_i of the i-th atom is rescaled accordingly, leading to an immediate rise in temperature model. A general form for the power density contains an amplitude factor S_0, an absorption law $d(x)$, an intensity distribution on the surface I(y,z), consideration of the reflectivity (1 - R), and the temporal beam shape $f(t)$:

$$S(x,t) = (1 - R) S_0 d(x) I(y,z) f(t) \tag{6.39}$$

For the absorption $d(x)$, the Lambert–Beer law is used, where α is the inverse absorption length. The intensity $I(y,z)$ can be adjusted according to the problem, while the temporal distribution $f(t)$ is assumed with a Gaussian shape:

$$S(x,t) = (1 - R) S_0 \alpha e^{-\alpha x} I(y,z) e^{-\frac{(t-t_0)^2}{2\sigma_t^2}} \tag{6.40}$$

The power density $S(x, t)$ is multiplied by the time step δt and is divided by the average atomic density ρn to achieve a small energy difference. After each time step δt, that amount of energy is added to the kinetic energy of the i-th particle:

$$E^i_{\text{kin}}(t + \delta t) = E^i_{\text{kin}}(t) + \frac{\delta t}{\rho_n} S(x_i, t) \tag{6.41}$$

The atomic number density ρ_n can be calculated on-the-fly by averaging the atomic masses over the irradiated volume. This density is assumed as constant for pulses less than 1000 fs. For significantly longer pulses, the material may change its phase, which results in a time-dependent density $\rho_n(t)$. The huge advantage of this model clearly lies in its simplicity: no knowledge of material properties such as electron–phonon coupling or processes during ablation is needed. Heat transfer, melting, and vaporization are all governed by the potentials used for the molecular dynamics.

6.2.2 *Results and Discussion*

The characteristics of laser plasma, as a nonlinear medium for HHG, play a crucial role in the restriction or enhancement of conversion efficiency. The growth of free electron density was likely the main limiting factor for the harmonic cut-off energy in early experiments with laser plasmas [79–81]. A search for appropriate materials, which can provide favorable ablation plasmas for efficient HHG, has motivated the analysis of plasma characteristics at conditions of high harmonic yield. Recent studies have shown that carbon ablation plasma is a prospective medium to satisfy the above requirements [73, 74, 82–84]. Other plasma plumes also demonstrate specific peculiarities of harmonic spectra, which are largely dependent on plasma consistency.

Firstly, we compared the results for HHG in carbon ablation plasmas and argon gas jets [85]. The pump laser intensity was in all cases 5×10^{14} W/cm^2 and duration 4.5 fs, pulse energy 2×10^{-4} J. Maximal harmonic order observed in carbon plasma was 27.

Our simulations were performed for two ablation times (8 ps and 10 ns) used in experiment [85] with the time-step of 1 fs. Laser field was treated as an addition to kinetic energy of every atom with respect to their depth $x$relative to sample surface. It should be noted that, for graphite, the reflection coefficient is relatively small and decreases with the increase of surface irregularity. So, it can safely be assumed as 0. The Brown module was considered as $\alpha = 0.01$ Å^{-1}. Sample surface depth was obtained from one-dimensional density estimate during each time-step. The positions and velocities of all carbon atoms were fixed every 50 fs. The velocity distribution was analyzed by counting the number of particles, which have the velocities sufficient to be in a given region ($\sim$200 μm above the ablated target) after 30 ns from the end of the pulse. Then, the number of particles was averaged for all measurements and divided by volume, which was the surface of the model sample multiplied by time and by difference between minimal and maximal speeds allowed.

For simulation of ablation by nanosecond pulses, we have to use the following approximate assumption. Firstly, the pulse intensity is considered constant and only total energy is the same as in experiments. Then, we perform ablation simulation directly for approximately 100 ps so that velocity distribution of ablated particles becomes almost constant. Finally, we extrapolate this velocity distribution for the whole duration of the pulse and get the particle concentration. We also used another approach of nanosecond pulses induced ablation simulation, which suggests $N = 10$ shorter simulations (5 ps) for unmodified sample, but every such simulation had a constant intensity corresponding to the intensity of the nanosecond pulse at N ns. However, the resulting concentration remained almost the same. We calculated the concentrations of carbon plasma at the experimental conditions of target ablation (i.e., 2×10^{10} W/cm^2 in the case of 8 ps pulses and 1×10^9 W/cm^2 in the case of 10 ns pulses). The corresponding concentrations were found to be 2.6×10^{17} cm^{-3} and 2.5×10^{18} cm^{-3}. The dependences of the plasma concentration on the intensities of 10 ns and 8 ps pulses at the ablating surface are summarized in Table 6.1.

In experimental studies [85], the density of the Ar gas jet used in comparative studies was measured by the absorption of XUV harmonics. Harmonics in the range of 30–60 eV were generated using a tube target filled with neon and propagated through the Ar gas jet that was placed 5 cm after the focus. Using known values of the Ar absorption cross-section in this energy range and assuming

Table 6.1 Calculations of carbon plasma concentration for different intensities of 8 ps and 10 ns pulses heating the graphite target

Intensity,	Plasma concentration, 10^{17} cm^{-3}	
10^9 W/cm^2	8 ps	10 ns
6.6	1.1	
20	2.6	
60	4.0	
0.33		10
1		25
3		37

a uniform gas distribution of 1.5 mm in length, an Ar density was estimated as 6×10^{17} cm^{-3} for the same operating conditions as used in the comparative study.

Experiments [85] have demonstrated the priority of carbon harmonics over argon harmonics in the case of low orders. This enhanced yield of carbon harmonics was explained by presence of nanoparticles generated in carbon plasma and their influence on the HHG conversion efficiency. One can expect that analogous behavior could occur in the case of other plasma plumes once the in situ nanoparticles formation follows the appropriate conditions of plasma formation (low concentration of free electrons, moderate excitation of ions, and so on).

Earlier, metal-ablating plasma characteristics were analyzed using the simulations by applying the thermodynamic code HYADES [86]. The HYADES code is a one-dimensional, Lagrangian hydrodynamics and energy-transport code. The electron and ion components are treated separately in a fluid approximation and are loosely coupled to each other, each in thermodynamic equilibrium. The diffusion approximation was used for modeling all energy-transport phenomena. The degree of ionization was determined by Saha, Thomas–Fermi, or other models.

The application of hydrodynamic code HYADES to determine the particle density of the carbon plume met with some uncertainties in absorbance of ablating target, although, in the case of metal targets, the values of plasma concentration have previously been reported in the range of 2×10^{17} cm^{-3} to 5×10^{17} cm^{-3}, which corresponded to the maximum conversion efficiency in the silver plasma [87].

To check the validity of our model, we also performed the MD simulations within our approach for similar conditions, but for widely used metal targets. These results, when directly compared with HYADES simulations of metal target ablation, show very good correspondence. This similarity can be in general explained by the fact that both approaches use the same simple absorption model and well-known absorption indexes of metals, so the amount of evaporated atoms should in general be the same if heat of vaporization is simulated properly, which is fulfilled by thermodynamic variables in HYADES and semi-empirical nature of inter-atomic potentials in our calculations.

Here, we briefly discuss the limitations of both approaches. Firstly, the thermodynamic approaches of HYADES are really fast and do not suffer much when increasing simulation time and delay. They also give reliable ionic concentrations, which could also be useful. On the other side, the choice of ablation targets is rather limited and complex molecules and alloys are not available in principle. Also, the thermodynamic calculations by nature consider the ablation of atomic particles only and have no support for cluster ablation; thus, in case of high intensities, they should be used with care. MD simulations, in their turn, can use any possible targets with no restrictions on its complexity provided correct potentials are given, which is a common task. They also can give an exact level of clusterization [88], but, unfortunately, for very short ablation pulses only due to discussed above quadratic complexity growth with time, which for exact clusterization computation cannot be avoided as we did. Most important, the code IMD ITAP is free and open-source. So, one can also successfully use generalization of short-time MD simulations for both main and reference targets as a correction to HYADES ablation results for the reference target (main target may not be available) and long simulation times.

Now let us discuss the influence of concentration on HHG saturation, which might take place in the obtained results and its connection to optimal plasma conditions. In experiments with this pump pulse, the cutoff energy for maximally observed 27th harmonic in carbon plasma was 46 eV $= 73.6 \times 10^{-19}$ J. So, if all the particles have this cutoff energy, only 2.71×10^{13} particles can participate in the HHG process. On the contrary, even for the generation of third harmonic, no more than 2.5×10^{14} interacting particles are needed (HHG full saturation boundary). One can assume that optimal plasma concentrations should be in this range.

For estimation of interacting volume of ablated carbon particles, we can represent it by an intersection of two cylinders, the first having the diameter of the driving beam in the focusing region ($R_1 = 0.3$ mm) and the second has the diameter of ablation spot ($R_2 = 0.6$ mm). The expression for this volume ($R_2 > R_1$) is

$$V(R_2, R_1) = \frac{8}{3}R_2\big[(R_2^2 + R_1^2)E(R_1/R_2) \\ -(R_2^2 - R_1^2)E(R_1/R_2)\big],\ R_2 > R_1 \qquad (6.42)$$

Here, $E(R_2/R_1)$ is the complete elliptic integral of the second kind [$E(0.5) = 1.47$] and the resulting volume is 5.28×10^{-5} cm^3. This sets the total number of interacting particles for 8 ps and 10 ns pulses to be 1.378×10^{13} and 1.32×10^{14}, respectively. We should clearly distinguish that particles are in general not necessary atoms, but ablation simulations give in fact concentration of atoms.

The interaction volume for argon is given by tube geometry and equals to 2.94×10^{-4} cm^3, which is almost five times higher than average carbon plasma volume. The total number of interacting argon particles is then 1.76×10^{14}. This number is very close to HHG saturation boundary, which has been, however, underestimated, as not all the passing radiation should necessarily be converted, and it has been previously noted that argon concentration was optimal as well (increasing the density of argon will only lead to lowering HHG yield in its high-energy part while having the same integral HHG yield). From this fact, we can deduce that our rough estimates of HHG saturation are in fact the key for determination of optimal plasma concentration for given pulse energy.

We can also make a very useful estimate of optimal concentration for carbon ablation plasmas, having in mind this interaction volume as a reference. The corresponding maximum nonrestricted cutoff harmonic energy (y) is plotted in Fig. 6.11 versus concentration (x). It is easily seen that HHG concentration saturation begins at particle concentration of 10^{17} cm^{-3} and becomes an important

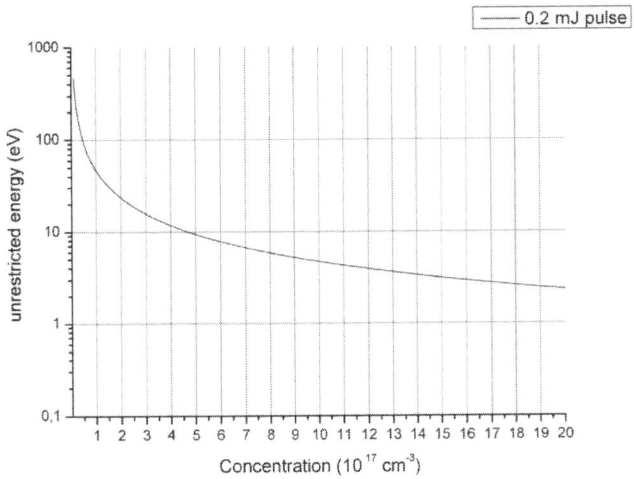

Figure 6.11 Unrestricted HHG energy dependence on particle concentration. Grid lines show that its influence is not extremely harmful for 10^{16}–5×10^{17} cm^{-3}. Pulse energy was 0.2 mJ.

factor at concentrations higher than 5×10^{17} cm^{-3}. We did not extrapolate these approaches on longer pulses because the energy for a given HHG act is absorbed and emitted only within one to two periods of the main pump, so not the whole pulse energy should be taken into consideration, leaving us with approximately the same considerations. Having compared these small estimates with optimal plasma ablation regimes and corresponding calculation of concentration, we can show that small nanoparticle formation is a way to get efficient HHG not only due to their enhanced nonlinear response but also to the overall decrease of single-atom particle concentration to optimal values (i.e., no more than 5×10^{17} cm^{-3}).

It has been shown that decrease of concentration to its optimal value is useful for increase of HHG yield for optimal carbon plasmas, because particle concentration of more than 10^{18} cm^{-3} (if it can be obtained at all during laser ablation) is indeed harmful for ablation plasma HHG experiments. Now let us check the difference between ablation regimes with the same ablation pulse energies. It should be noted that our simulation due to its approximated piecewise nature showed only the concentration of carbon atoms in interaction region, not the shape of ablated particles. It can be supposed that nanoparticle formation is the most probable effect during ablation, because carbon atoms due to their high reactivity tend to form nanostructures while in plasma phase, and the higher is atom concentration, the higher is nanoparticle formation rate. On the contrary, short intense ablation by 8 ps pulses can result in chipping of very large pieces of graphite, which absorb pump radiation much more effectively than atoms and small nanoparticles, but only break to atoms without HHG generation in the process. So, the 10 ns pulses are preferable not by giving optimal concentration, but first of all by the possibility of more uniform and controllable ablation of atoms, which could lead to small nanoparticles formation.

Previous studies have demonstrated that, during 8 ps pulses ablation, very large particles have been obtained, but for 10 ns regime, they have not been noticed unless fluency has been increased [83, 85]. However, mass spectroscopy scan of 10 ns pulses-induced ablated plasma has shown that there are indeed very small nanoparticles (from 10 to 30 particles in a cluster). Carbon nanoparticles are known for their possibility to enhance HHG yield due to local-field effects, which describes better efficiency of longer ablation pulses and increased HHG yield related to metal plasma experiments. The discussed studies have demonstrated that computation of optimal plasma concentration is now possible on a very reliable basis of molecular dynamical calculations, which could be further extended toward the nanoparticles formation.

6.3 Comparison of High-Order Harmonic Generation in Uracil and Thymine Ablation Plumes

Previous studies of laser plasma HHG have been mostly confined to ablation targets constituted by a single element, such as graphite or silver, or by two-element targets, for example, gallium arsenide. The extension of the technique to more complex species, including biological molecules, is of high interest, as these studies may help unravel some of the ultrafast electronic dynamics that take part in processes such as deoxyribonucleic acid (DNA) damage. To approach the application of laser plasma HHG spectroscopy to large molecules, we have examined in this work the ribonucleic acid

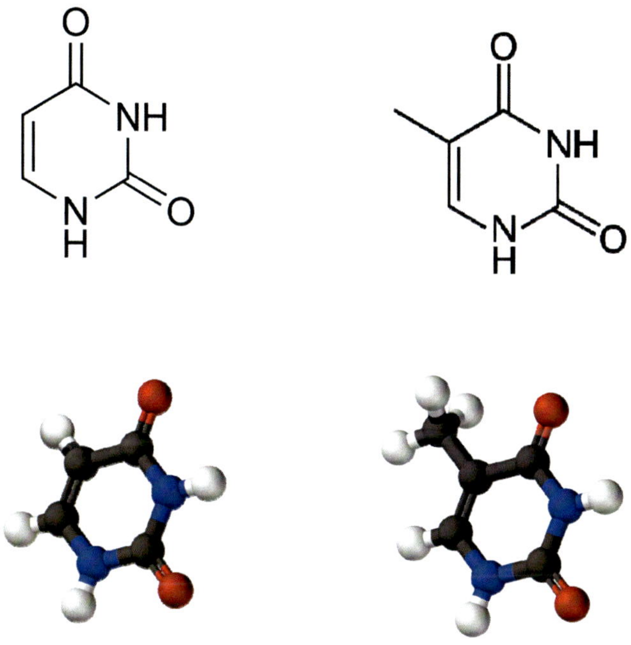

Figure 6.12 Molecular (top) and orientational (bottom) structures of uracil (left column) and thymine (right column). Reproduced from Ref. 89 with permission from the PCCP Owners Society.

(RNA) base uracil and have established a comparison with its DNA counterpart, thymine [89]. Uracil has been successfully vaporized in order to be studied [90]; however, the gaseous densities achievable by such methods are low. For HHG studies, we require a relatively high density (10^{15}–$10^{17}\,cm^{-3}$) of emitters, and for quantitative HHG spectroscopy, there must be intact molecules in gas phase. The objective is to generate high density, weakly ionized, ablation plumes composed predominately of un-fragmented parent molecules.

Uracil ($C_4H_4N_2O_2$) is a naturally occurring pyrimidine derivative and one of the four nucleobases in RNA in which uracil binds to adenine via two hydrogen bonds. Methylation of uracil produces thymine ($C_5H_6N_2O_2$), which replaces uracil in DNA. The similar molecular structures of uracil and thymine are shown in Fig. 6.12. At room temperature, both uracil and thymine are solids with melting points of 335°C and 317°C, respectively. The low-lying vertical ionization potential of uracil is 9.5 eV while that of thymine is 9.1 eV [91]. Thus, with similar chemical and physical properties, we might expect to see similar non-linear optical properties of the ablation plume formed from these compounds.

In this section, I describe HHG from laser ablation plasmas of uracil and thymine targets using heating pulses of 1064 nm, 10 ns and of 780 nm, 160 ps lasers. Harmonics were observed from uracil plasma using both 780 nm and 1300 nm driving radiation with pulses of 30 fs under the two laser ablation regimes. In contrast, the nonlinear optical response of the thymine plume appeared to be quite different, as no signs of HHG were observed. We studied the composition of the plumes using time of flight mass spectrometry (TOFMS) and examined the plume debris by scanning electron microscopy (SEM). These investigations suggest that ablation in these conditions leads to a higher atomization degree in the case of thymine, while in the case of uracil, larger fragments and intact parent molecules survive the process. These studies constitute the first attempt to analyze differences in structural properties of complex molecules through plasma ablation induced HHG spectroscopy.

6.3.1 *Experiment*

A 1 kHz repetition rate, chirped pulse amplification Ti:sapphire laser was used as the main laser driving source. A part of the uncompressed CPA radiation was split off from the beam line prior to the compressor stage (780 nm, 1.3 mJ, 160 ps) and used for ablation of the solid targets placed in a vacuum chamber. This beam was loosely focused onto the target with a spot size of about 500 μm. The resulting intensity on the target surface, of 2×10^{10} W/cm^2, created a weakly ionized plasma plume. As an alternative ablation laser, we also used the 10 ns, 1064 nm pulses from a 10 Hz repetition rate Q-switched Nd:YAG laser that provided an intensity on the target surface of 1×10^9 W/cm^2. In that case, the delay between the 10 ns heating pulse and the 30 fs probe pulse was varied in the range of 10-80 ns to maximize the harmonic yield. A more detailed description of the setup can be found in [92]. After a delay of nearly 60 ns, required for formation and expansion of the plasma plume away from the surface of the target, the compressed laser pulse (780 nm, 0.6 mJ, 30 fs) was focused onto the plasma using a 200 mm focal length mirror up to an intensity of 1×10^{15} W/cm^2 to generate the high-order harmonics. The position of the focus relative to the plume was chosen to maximize harmonic generation efficiency.

HHG experiments were also performed using mid-infrared (MIR) radiation as a driving laser source. A commercial optical parametric amplifier (OPA) was pumped with 7 mJ, 30 fs, 780 nm radiation to generate the MIR pulses. This system produced 40 fs pulses with tunable wavelength between 1200 nm and 1550 nm in the signal wave with a maximum energy at 1300 nm. The idler pulse covered the 1650–2200 nm range with maximum energy at 2000 nm. The harmonic emission was analyzed using an extreme ultraviolet spectrometer consisting of a flat field grating and micro channel plate coupled to a phosphor screen. Images of harmonics were captured using a CCD camera.

Uracil and thymine (>99.0% purity) targets were prepared by stamping the powdered materials into solid pellets using a die and fly press. This method ensured a smooth surface to be exposed to laser ablation. For some experiments, targets containing 50:50 weight ratio of both compounds were also prepared. The targets were rotated to minimize overheating and damage of their surface from repeated laser shots, thus ensuring more stable ablation conditions [93]. Characterization of the plasma debris collected on silicon wafers, placed 4 cm from the ablated target, was carried out by SEM.

Ablation of powdered compounds is susceptible to shot-to-shot instability and rapid decrease of the harmonic yield in the case of static samples, due to the abrupt change in target morphology following ablation. Such instabilities limit the application of powdered materials for plasma HHG spectroscopy [94]. In the discussed studies, we implemented the use of rotating targets in order to improve harmonic stability from plasma plumes produced from the compressed powder. We found that once the target rotation is stopped, the harmonic efficiency from the plasma decreased to the noise level within 1–2 seconds.

The nanosecond laser ablation plume was investigated by TOFMS. A brief summary of this method is given here (see [95] for a full description). The laser beam (1064 nm, 10 ns pulse duration) was focused on the surface of the uracil/thymine target, at normal incidence. The ablation fluence employed for the measurements matched the range used in the HHG experiments using nanosecond ablation. The target was placed in a vacuum chamber (pumped to $\sim 2\times10^{-6}$ mbar) between the extracting and accelerating plates of a linear TOFMS. The target surface was parallel to the flight axis of the spectrometer and could be rotated and displaced at variable distances from the axis. Positive ions created during the ablation process were deflected along the TOFMS axis by an electric field typically in the range of 300–400 V cm^{-1} and accelerated by a total voltage of nearly 2500 V. Ions entered the drift region (flight length $\sim$1 m) and were detected by a set of spatially integrating microchannel plates.

6.3.2 *Results and Discussion*

Harmonics were generated from the uracil plasma plume using the driving laser pulses at wavelengths of 780 nm and 1300 nm (Fig. 6.13). One can see that the harmonics generated at the two wavelengths are of comparable intensity, even if the energy in the 1300 nm pulse is lower than that of the 780 nm pulses (0.35 mJ and 0.53 mJ, respectively). This pattern of intensities does not follow the usual wavelength scaling rule for atoms of harmonic intensity $I_h \propto \lambda^{-5}$ [96, 97], which, for this case, would predict a nearly 13-fold decrease in conversion efficiency for the MIR source at equal experimental conditions.

We compared uracil with its DNA equivalent thymine in various sets of experiments. The results in the case of 780 nm driving pulses are presented in Fig. 6.14. The pure uracil spectra (bottom panel) clearly showed high-order harmonics, although no such signal was detected from the thymine plume (upper panel). Despite persistent attempts at different ablation intensities, using both the ns and ps ablation pulses and delays, we were unable to observe any HHG signal from the thymine ablation plume. The 50:50 mixture of the two compounds (middle panel) displayed the same harmonic spectra as pure uracil, although with lower intensity, as expected, given that thymine plume species do not seem to participate in harmonic generation.

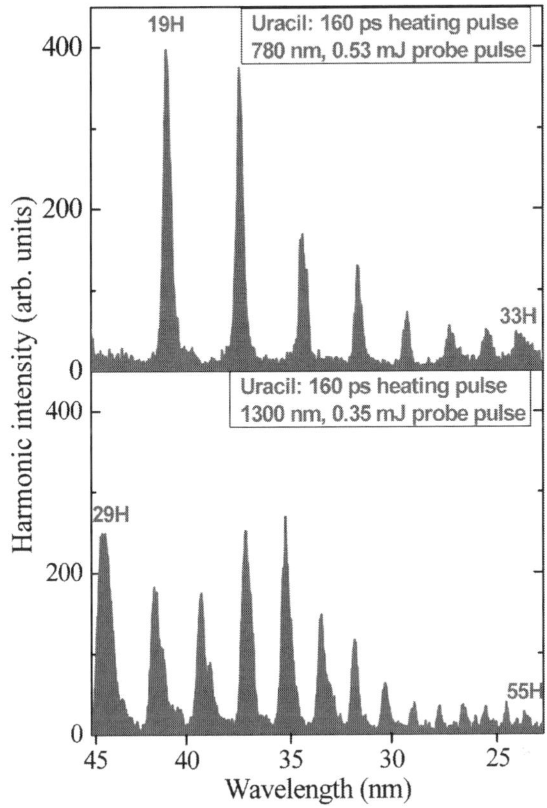

Figure 6.13 HHG in uracil plasma plume using 780 and 1300 nm probe pulses and 780 nm, 160 ps ablating pulses. Reproduced from Ref. 89 with permission from the PCCP Owners Society.

From the above, it is clear that there are differences in the non-linear response of the ablation plumes of the two pyrimidine nucleobases uracil and thymine. For better understanding of the observed differences, the composition of the plumes was analyzed via TOFMS. The mass spectra collected on uracil and thymine nanosecond laser ablation plumes are presented in Fig. 6.15. The spectrum of uracil shows the presence of C, C_2, C_3, and larger molecular fragment ions together with a small peak at 112 amu, which is assigned to the uracil parent ion. The signal in the TOFMS of the thymine plume is shifted toward lower masses and larger

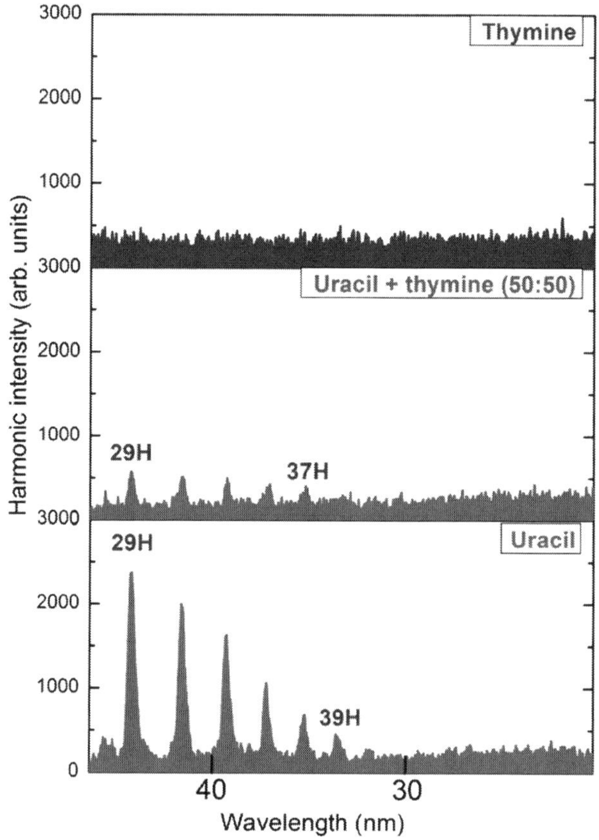

Figure 6.14 XUV spectra from uracil, thymine and a target made of a 50:50 mixture of the uracil and thymine in the case of 1300 nm probe pulses and 780 nm, 160 ps ablating pulses ($I = 2\times 10^{10}$ W/cm^2). Reproduced from Ref. 89 with permission from the PCCP Owners Society.

fragments and parent ions peaks are negligible. Thymine shows a far more complete breakup, predominately into C^+ peak. This may indicate a significantly higher free electron density as well as a strong depletion of parent molecular species in the plume. To gain further insight into the composition of the plume, the debris produced upon ablation of the targets for a fixed period of time was collected on silicon plates placed in close proximity and then analyzed by SEM. As an example, Fig. 6.16 shows images of deposits

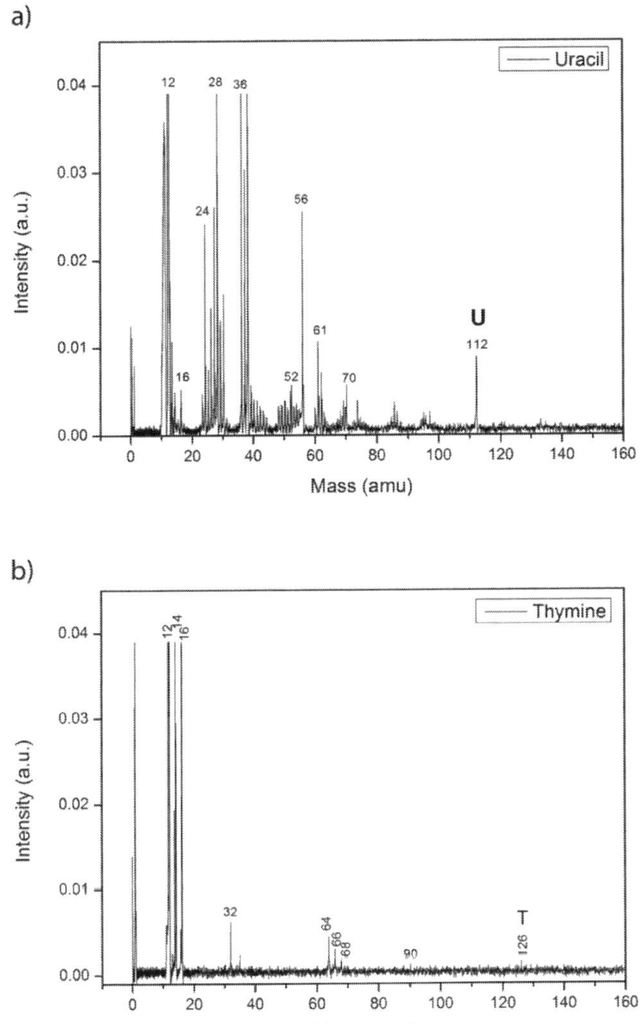

Figure 6.15 Time of flight mass spectra showing the ion composition of uracil (a) and thymine (b) ablation plumes created by 1064 nm 10 ns pulses at a fluence similar to that used in HHG experiments. The parent ions for uracil and thymine are marked as U and T. Reproduced from Ref. 89 with permission from the PCCP Owners Society.

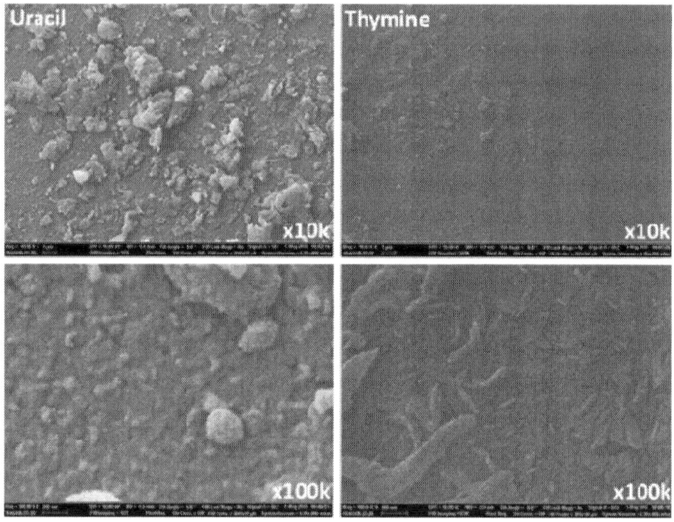

Figure 6.16 SEM images of deposits collected from uracil (left column) and thymine (right column) targets upon ablation during 10 min. with 10 ns, 1064 nm pulses at 10 Hz. Reproduced from Ref. 89 with permission from the PCCP Owners Society.

collected during 10 min upon ablation at 10 Hz with the 1064 nm, 10 ns pulses. A higher density of particulates is observed in uracil, as compared with thymine, suggesting that, either the former gives rise to denser plasmas under the same laser ablation intensity, or that the plume species reaching the silicon substrate tend to aggregate more efficiently. It is also observed that deposits from the uracil plasma display a large number of spheroidal particles with relatively similar size (~40–100 nm), whereas in contrast, thymine deposits present a crystalline-like structure.

These studies constitute the first attempt to analyze the influence of structural properties on the nonlinear response of similar molecular compounds, which only differ by the methylation of one of the ring carbon atoms, through plasma ablation-induced HHG spectroscopy. Although in the case of uracil efficient HHG was observed in the ablation plume of this compound, we did not observe HHG from thymine in a wide range of explored experimental conditions.

As already mentioned, there are large differences in ion composition of the respective ablation plumes. We believe this to be the most likely candidate for the strongly dissimilar nonlinear behavior during the interaction with the strong driving laser field. Irradiation of the solid target causes excitation and eventually ejection of the parent molecule. Molecular fragments and clusters are also produced upon direct interaction of the ablating pulse with the target material and by collisions in the expanding plume. Excited state dynamics of uracil and thymine [98, 99] is understood in terms of rapid internal conversion mediated by conical intersections. The $^1(\pi\pi^*)$ HOMO–LUMO singlet excited molecular state lies at 5.1 and 4.8 eV above the ground state, with oscillator strengths of 0.142 and 0.147 in uracil and thymine, respectively [100]. The likely multiphoton absorption process, which couples the ablating pulse energy to the targets, should result in a similar ablation threshold for the two compounds. The discussed results have shown that thymine has a 20% larger ablation threshold when compared with uracil, therefore, requiring larger pulse energy in order to achieve a similarly high density in the case of thymine, which could lead to greater fragmentation. The resulting photochemistry and photofragmentation patterns could present particularities connected to the molecular structure. Both uracil and thymine have a similar strong tendency to photodimerization [101] and both molecules are likely to form dimers through hydrogen bonding. We argue that the differences in the ablation plumes of uracil and thymine are correlated with different fragmentation patterns of the parent molecules in the plume.

As shown for nanosecond ablation, the uracil plume is richer than the thymine plume in carbon species and larger molecular fragments and it also contains surviving intact parent molecular ions, which are almost entirely absent in the case of thymine. Aggregation in the plasma plume will favor cluster and nanoparticles formation that act as efficient nonlinear species, explaining the efficient high-frequency up-conversion and the higher density of deposits of nanometer dimensions in the case of uracil. In contrast, the plume of thymine is richer in atomic carbon and low mass species. The higher degree of atomization of the thymine plume will favor the growth of crystalline deposits as observed by SEM (Fig. 6.16) [102, 103]. In addition, the higher degree of atomization in the thymine plume (appearing as atomic ions) will lead to a higher density of plasma-free electrons. Free electrons will contribute abundantly in the case of thymine to phase mismatch, which is known to have a clear detrimental effect upon HHG efficiency [104].

Summarizing this section, we have discussed the first studies of the HHG of laser radiation from laser ablation plumes of the RNA nucleobase uracil. Harmonics were observed using both 30 fs, 780 nm and 1300 nm fundamental laser radiation from the plumes generated upon ablation of solid targets with 1064 nm, 10 ns and 780 nm, 160 ps pulses. Contrary to uracil, HHG was not observed in ablation plumes of its DNA counterpart thymine under any of the explored conditions. We believe that the considerable difference in the nonlinear optical response is related to differences in chemical structure of the two pyrimidine nucleobases and to different composition of their respective ablation plumes. The fragmentation patterns induced upon laser ablation in each compound are markedly different despite their relatively similar molecular structure and give rise to a characteristic distribution of plasma species in the ablation plumes showing a higher degree of fragmentation in the case of thymine. The uracil plume is richer in larger molecular carbon based fragments; these species tend to aggregate resulting in a larger nonlinear response. In contrast, the plume of thymine contains a higher density of atomic and low molecular weight fragment ions and consequently gives rise to a higher electronic density, which contributes to phase mismatch and eventually leads to a negligible high harmonic signal. It is possible that by judicious control of the laser ablation conditions, plumes with high content of parent molecular ions could be formed. Finding those conditions, if they exist, remains a challenge and requires further investigation. These studies constitute the first attempt to analyze differences in structural molecular properties through plasma ablation induced HHG spectroscopy and show the potential of the technique for understanding structural as geometrical parameters of medium-size organic molecules of biological interest.

References

1. A. V. Andreev, R. A. Ganeev, H. Kuroda, S. Y. Stremoukhov, O. A. Shoutova, *Eur. Phys. J. D* 67, 22 (2013).
2. N. A. Papadogiannis, L. A. A. Nikolopoulos, D. Charalambidis, *Phys. Rev. Lett.* 90, 133902 (2003).
3. J. Itatani, J. Levesque, D. Zeidler, H. Niikura, H. Pépin, J. C. Kieffer, P. B. Corkum, D. M. Villeneuve, *Nature* 432, 867 (2004).
4. R. I. Tobey, M. E. Siemens, O. Cohen, M. M. Murnane, H. C. Kapteyn, K. A. Nelson, *Opt. Lett.* 32, 286 (2007).
5. M. Drescher, M. Hentschel, R. Kienberger, G. Tempea, C. Spielmann, G. A. Reider, P. B. Corkum, F. Krausz, *Science* 291, 1923 (2001).
6. P. B. Corkum, F. Krausz, *Nature Phys.* 3, 381 (2007).
7. P. V. Redkin, R. A. Ganeev, *Phys. Rev. A* 81, 063825 (2010).
8. R. A. Ganeev, H. Singhal, P. A. Naik, U. Chakravarty, V. Arora, J. A. Chakera, R. A. Khan, M. Raghuramaiah, S. R. Kumbhare, R. P. Kushwaha, P. D. Gupta, Appl. Phys. B 87, 243 (2007).
9. R. A. Ganeev, Bom L. B. Elouga, J.-C. Kieffer, T. Ozaki, *Phys. Rev. A* 76, 023831 (2007).
10. R. A. Ganeev, M. Baba, M. Suzuki, H. Kuroda, *Phys. Lett. A* 339, 103 (2005).
11. R. A. Ganeev, *Phys. Uspekhi* 52, 55 (2009).
12. M. Suzuki, M. Baba, R. A. Ganeev, H. Kuroda, T. Ozaki, *Opt. Express* 15, 1161 (2007).
13. L. V. Keldysh, *Sov. Phys. JETP* 20, 1307 (1965).
14. A. M. Perelomov, V. S. Popov, M. V. Terent'ev, *Sov. Phys. JETP* 24, 207 (1967).
15. M. Perelomov, V. S. Popov, *Sov. Phys. JETP* 25, 336 (1967).
16. V. S. Popov, V. P. Kuznetsov, A. M. Perelomov, *JETP* 26, 222 (1968).
17. F. H. M. Faisal, *J. Phys. B* 6, L89 (1973).
18. H. R. Reiss, *Phys. Rev. A* 22, 1786 (1980).
19. H. R. Reiss, *Phys. Rev. A* 42, 1476 (1990).
20. H. R. Reiss, *Prog. Quantum Electron* 16, 1 (1992).
21. M. Pont, M. Gavrila, *Phys. Rev. Lett.* 65, 2362 (1990).
22. K. C. Kulander, K. J. Schafer, J. L. Krause, *Phys. Rev. Lett.* 66, 2601 (1991).
23. P. Marte, P. Zoller, *Phys. Rev. A* 43, 1512 (1991).
24. W. C. Henneberger, *Phys. Rev. Lett.* 21, 838 (1968).

25. M. Gavrila, J. Kaminski, *Phys. Rev. Lett.* 52, 613 (1984).
26. E. A. Volkova, A. M. Popov, O. V. Tikhonova, *JETP* 93, 1155 (2001).
27. W. Becker, S. Long, J. K. McIver, *Phys. Rev. A* 41, 4112 (1990).
28. W. Becker, A. Lohr, M. J. Kleber, *Phys. B* 27, L325 (1994).
29. S. Long, W. Becker, J. K. McIver, *Phys. Rev. A* 52, 2262 (1995).
30. G. Bandarage, A. Maquet, T. Mienis, R. Taieb, J. Cooper, *Phys. Rev. A* 46, 380 (1992).
31. A. V. Andreev, S. Y. Stremoukhov, O. A. Shoutova, *J. Russian Laser Res.* 29, 203 (2008).
32. K. C. Kulander, K. J. Schafer, J. L. Krause, *Atoms in Intense Laser Fields,* ed. M. Gavrila, Academic Press, New York (1992).
33. M. Protopapas, C. H. Keitel, P. L. Knight, *Rep. Prog. Phys.* 60, 389 (1997).
34. J. H. Eberly, Q. Su, J. Javanainen, *Phys. Rev. Lett.* 62, 881 (1989).
35. K. C. Kulander, *Phys. Rev. A* 35, R445 (1987).
36. K. C. Kulander, *Phys. Rev. A* 36, 2726 (1987).
37. P. L. DeVries, *J. Opt. Soc. Am. B* 7, 517 (1990).
38. K. J. LaGutta, *J. Opt. Soc. Am. B* 7, 639 (1990.
39. M. Y. Emelin, M. Y. Ryabikin, A. M. Sergeev, *JETP* 133, 243 (2008).
40. V. V. Strelkov, A. A. Gonoskov, I. A. Gonoskov, M. Y. Ryabikin, *Phys. Rev. Lett.* 107, 043902 (2011).
41. J. L. Krause, K. J. Schafer, K. C. Kulander, *Phys. Rev. Lett.* 68, 3535 (1992).
42. P. Antoine, B. Piraux, A. Maquet, *Phys. Rev. A* 51, R1750 (1995).
43. P. Antoine, D. B. Milošević, B. Piraux, M. Gajda, *Laser Phys.* 7, 3 (1997).
44. E. Cormier, P. Lambropoulos, *J. Phys. B* 29, 1667 (1996).
45. E. Cormier and P. Lambropoulos, *J. Phys. B* 30, 77 (1997).
46. D. Bauer, D. B. Milošević, W. Becker, *Phys. Rev. A* 72, 023415 (2005).
47. M. Lewenstein, P. Balcou, M. Y. Ivanov, A. L'Huillier, P. B. Corcum, *Phys. Rev. A*, 49, 2117 (1994).
48. M. Krug, T. Bayer, M. Wollenhaupt, C. Sarpe-Tudoran, T. Baumert, S. S. Ivanov, M. V. Vitanov, *New J. Phys.* 11, 105051 (2009).
49. A. V. Andreev, S. Y. Stremoukhov, O. A. Shoutova, *JETP* 111, 6 936 (2010).
50. X. Tang, H. Rudolph, P. Lambropoulos, *Phys. Rev. A* 44, R6994 (1991).
51. H. Xu, X. Tang, P. Lambropoulos, *Phys. Rev. A* 46, R2225 (1992).
52. L. A. A. Nikolopoulos, P. Lambropoulos, *Phys. Rev. A* 82, 3771 (1999).
53. A. V. Andreev, *JETP* 89, 421 (1999).

54. A. V. Andreev, S. Yu. Stremoukhov, O. A. Shoutova, *JETP Lett.* 93, 476 (2011).
55. A. V. Andreev, S. Yu. Stremoukhov, O. A. Shoutova, *Laser Phys.* 17, 496 (2007).
56. A. V. Andreev, S. Yu. Stremoukhov, O. A. Shoutova, *Eur. Phys. J. D* 66, 1245 (2012).
57. R. A. Ganeev, *J. Phys. B: At. Mol. Opt. Phys.* 40, R213 (2007).
58. M. Goppert-Mayer, *Ann. der Phys.* 9, 273 (1931).
59. C. Hutchison, R. A. Ganeev, T. Witting, F. Frank, W. A. Okell, J. W. G. Tisch, J. P. Marangos, *Opt. Lett.* 37, 2064 (2012).
60. L. D. Landau, E. M. Lifshits, *Quantum Mechanics*, 3rd ed., Saint Louis, Elsevier Science (1981).
61. H. Bethe, E. Salpeter, *Quantum Mechanics of One- and Two-Electron Atoms, Berlin*, Academic Press Inc. (1957).
62. P. B. Corkum, *Phys. Rev. Lett.* 71, 1994 (1993).
63. Winterfeldt, C. Spielmann, G. Gerber, *Rev. Mod. Phys.* 80, 117 (2008).
64. T. Popmintchev, M-Ch. Chen, P. Arpin, M. M. Murnane, H. C. Kapteyn, *Nature Phot.* 4, 822 (2010).
65. T. D. Donnelly, T. Ditmire, K. Neuman, M. D. Perry, R. W. Falcone, *Phys. Rev. Lett.* 76, 2472 (1996).
66. J. W. G. Tisch, T. Ditmire, D. J. Fraser, N. Hay, M. B. Mason, E. Springate, J. P. Marangos, M. H. R. Hutchinson, *J. Phys. B: At. Mol. Opt. Phys.* 30, L709 (1997).
67. R. A. Ganeev M. Suzuki, M. Baba, M. Ichihara, H. Kuroda, *J. Appl. Phys.* 103, 063102 (2008).
68. H. Singhal, R. A. Ganeev, P. A. Naik, A. K. Srivastava, A. Singh, R. Chari, R. A. Khan, J. A. Chakera, P. D. Gupta, *J. Phys. B: At. Mol. Opt. Phys.* 43, 025603 (2010).
69. M. Ferray, A. L'Huillier, X. F. Li, L. A. Lompre, G. Mainfray, C. Manus, *J. Phys. B At. Mol. Opt. Phys.* 21, L31 (1988).
70. A. Pirri, C. Corsi, M. Bellini, *Phys. Rev. A* 78, 011801 (2008).
71. R. A. Ganeev, M. Suzuki, T. Ozaki, M. Baba, H. Kuroda, *Opt. Lett.* 31, 1699 (2006).
72. J.-F. Hergott, M. Kovacev, H. Merdji, C. Hubert, Y. Mairesse, E. Jean, P. Breger, P. Agostini, B. Carré, P. Salières, *Phys. Rev. A* 66, 021801 (2002).
73. L. B. Elouga Bom, Y. Pertot, V. R. Bhardwaj, T. Ozaki, *Opt. Express* 19, 3077 (2011).

74. Y. Pertot, L. B. Elouga Bom, V. R. Bhardwaj, T. Ozaki, *Appl. Phys. Lett.* 98, 101104 (2011).
75. R. A. Ganeev, L. B. Elouga Bom, J. Abdul-Hadi, M. C. H. Wong, J. P. Brichta, V. R. Bhardwaj, T. Ozaki, *Phys. Rev. Lett.* 102, 013903 (2009).
76. R. A. Ganeev, P. A. Naik, H. Singhal, J. A. Chakera, M. Kumar, M. P. Joshi, A. K. Srivastava, P. D. Gupta, *Phys. Rev. A* 83, 013820 (2011).
77. J. Tersoff, *Phys Rev. B* 37, 6991 (1988).
78. J. Tersoff, *Phys. Rev. Lett.* 61, 2879 (1988).
79. Y. Akiyama, K. Midorikawa, Y. Matsunawa, Y. Nagata, M. Obara, H. Tashiro, K. Toyoda, *Phys. Rev. Lett.* 69, 2176 (1992).
80. C.-G. Wahlström, S. Borgström, J. Larsson, S.-G. Petterson, *Phys. Rev. A* 51, 585 (1995).
81. W. Theobald, C. Wülker, F. R. Schäfer, B. N. Chichkov, *Opt. Commun.* 120, 177 (1995).
82. R. A. Ganeev, P. A. Naik, J. A. Chakera, H. Singhal, N. C. Pramanik, P. A. Abraham, N. Rani Panicker, M. Kumar, P. D. Gupta, *J. Opt. Soc. Am. B* 28, 360 (2011).
83. R. A. Ganeev, T. Witting, C. Hutchison, F. Frank, P. V. Redkin, W. A. Okell, D. Y. Lei, T. Roschuk, S. A. Maier, J. P. Marangos, J. W. G. Tisch, *Phys. Rev. A* 85, 015807 (2012).
84. Y. Pertot, S. Chen, S. D. Khan, L. B. Elouga Bom, T. Ozaki, Z. Chang, *J. Phys. B: At. Mol. Opt. Phys.* 45, 074017 (2012).
85. R. A. Ganeev, C. Hutchison, T. Witting, F. Frank, W. A. Okell, A. Zaïr, S. Weber, P. V. Redkin, D. Y. Lei, T. Roschuk, S. A. Maier, I. López-Quintás, M. Martín, M. Castillejo, J. W. G. Tisch, J. P. Marangos, *J. Phys. B: At. Mol. At. Phys.* 45, 165402 (2012).
86. M. Rubenchik, M. D. Feit, M. D. Perry, J. T. Larsen, *Appl. Surf. Sci.* 129, 193 (1998).
87. L. B. Elouga Bom, J.-C. Kieffer, R. A. Ganeev, M. Suzuki, H. Kuroda, T. Ozaki, *Phys. Rev. A* 75, 033804 (2007).
88. J. Roth, C. Trichet, H.-R. Trebin, S. Sonntag, in *High performance computing in science and engineering '10*, W. E. Nagel, D. B. Kröner and M. M. Resch, eds., Springer, Berlin (2011).
89. C. Hutchison, R. A. Ganeev, M. Castillejo, I. Lopez-Quintas, A. Zair, S. J. Weber, F. McGrath, Z. Abdelrahman, M. Oppermann, M Martín, D. Y. Lei, S. A. Maier, J. W. Tisch, J. P. Marangos, *Phys. Chem. Chem. Phys.* 15, 12308 (2013).

90. V. Vaquero, M. E. Sanz, J. C. López, J. L. Alonso, *J. Phys. Chem. A* 111, 3443 (2007).
91. D. Roca-Sanjuán, M. Rubio, M. Merchán, L. Serrano-Andrés, *J. Chem. Phys.* 125, 084302 (2006).
92. R. A. Ganeev, V. V. Strelkov, C. Hutchison, A. Zaïr, D. Kilbane, M. A. Khokhlova, J. P. Marangos, *Phys. Rev. A* 85, 023832 (2012).
93. C. Hutchison, R. A. Ganeev, T. Witting, F. Frank, W. A. Okell, J. W. G. Tisch, J. P. Marangos, *Opt. Lett.* 37, 2064 (2012).
94. R. A. Ganeev, L. B. Elouga Bom, J. Abdul-Hadi, M. C. H. Wong, J. P. Brichta, V. R. Bhardwaj, T. Ozaki, *Phys. Rev. Lett.* 102, 013903 (2009).
95. R. Torres, M. Jadraque, M. Martin, *Appl. Phys. A* 80, 1671 (2005).
96. J. Tate, T. Auguste, H. G. Muller, P. Salières, P. Agostini, L. F. DiMauro, *Phys. Rev. Lett.* 98, 013901 (2007).
97. K. Schiessl, K. L. Ishikawa, E. Persson, J. Burgdörfer, *Phys. Rev. Lett.* 99, 253903 (2007).
98. M. Merchán, R. González-Luque, T. Climent, L. Serrano-Andrés, E. Rodríguez, M. Reguero, D. Peláez, *J. Phys. Chem. B* 110, 26471 (2006).
99. M. Schneider, C. Schon, I. Fischer, L. Rubio-Lago, T. Kitsopoulos, *Phys. Chem. Chem. Phys.* 9, 6021 (2007).
100. T. Gustavsson et. al. *J. Am. Chem. Soc.* 128, 607 (2006).
101. K. C. Smith, *Photochem. Photobiol.* 2, 503 (1963).
102. K. Ozeki, N. Sakabe, J. Tanaka, *Acta Cryst. B* 25, 1038 (1969).
103. G. S. Parry, *Acta Cryst.* 7, 313 (1954).
104. M. Schnürer, Z. Cheng, M. Hentschel, G. Tempea, P. Kálmán, T. Brabec, F. Krausz, *Phys. Rev. Lett.* 83, 722 (1999).

Chapter 7

Recent Achievements in Plasma Harmonics

In this chapter, I show some most recent developments in plasma HHG. In particular, I present the results of experimental studies of HHG in plasmas containing fullerenes using pulses of different duration and wavelength. We find that the harmonic cut-off is extended in the case of few-cycle pulses (3.5 fs, 29th harmonic) compared with longer laser pulses (40 fs, 25th harmonic) at the same intensity. Our studies also include HHG in fullerenes using 1300 and 780 nm multi-cycle (35 and 40 fs) pulses. For 1300 nm pulses, an extension of the harmonic cutoff to the 41st order was obtained, with a decrease in conversion efficiency that is consistent with theoretical predictions based on wavepacket spreading for single atoms.

Next, I present an extension of single active electron (SAE) rigid sphere fullerene paradigm, which now includes ionization and radial movement of electron in intense laser field. We compare the theoretical results with experimental high-order harmonic generation (HHG) in fullerenes.

Plasma Harmonics
Rashid A. Ganeev

ISBN 978-981-4463-30-0 (Hardcover), 978-981-4463-31-7 (eBook)
www.panstanford.com

Further, I discuss first studies of the dynamics of nanoparticles ablation and HHG in nanoparticle-containing plasmas at 1 kHz pulse repetition rate using rotation target technique. Analysis of the morphology of ablated clusters allowed defining the conditions that lead to the generation of a plasma, which contains nanoparticles of the same characteristics as the ones used as target material. Application of these plasma plumes as the media for harmonic generation of laser radiation allowed the efficient conversion of the IR photons toward the short-wavelength range. Comparison with harmonics generated in ablation plasma of bulk targets of the same metallic element clearly demonstrates the superior efficiency of the nanoparticle-based targets. The comparative studies of harmonic generation from single-particle and multiparticle plasmas showed the advantages in the latter case. We discuss the mechanisms of efficient harmonic generation in cluster-containing media.

7.1 High-Order Harmonic Generation in Fullerenes Using Few and Multi-Cycle Pulses of Different Wavelengths

The ionization saturation intensities of different charge states of C_{60} are higher than isolated atoms of similar ionization potential [1]. With this perspective, it is interesting to analyze HHG from fullerene molecules in the field of few-cycle laser pulses and compare these studies with those carried out using multi-cycle pulses. The motivation for the present work was to analyze the conditions for efficient HHG from plasmas containing C_{60}, using picosecond laser pulses to ablate the fullerene-containing target at high pulse repetition rate (1 kHz) and then few-cycle laser pulses ($\tau = 3.5$ fs) to generate the harmonics in the fullerene plasma. We also investigated HHG in fullerenes using 1300 nm radiation and compared these studies with those using 780 nm multi-cycle pulses. We employed a new technique that led to a considerable improvement in the stability of the harmonics from the plasma produced on the fullerene-containing targets, which is crucial in high pulse repetition rate experiments [2].

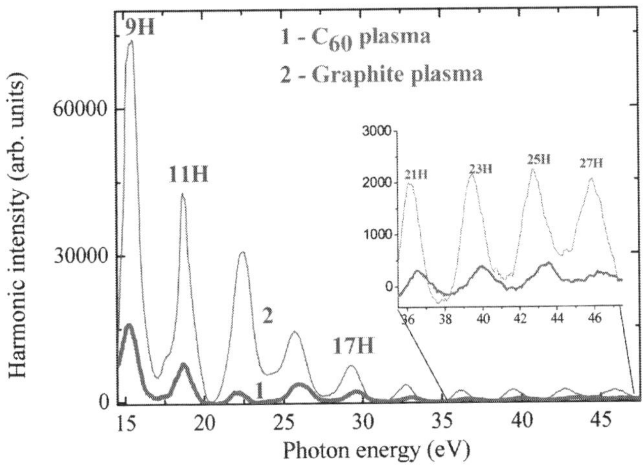

Figure 7.1 Harmonic spectra from fullerene (thick curve) and graphite (thin curve) plasmas using the heating 8 ps pulses and 3.5 fs probe pulses at identical experimental conditions. Reproduced from Ref. 2 with permission from Optical Society of America.

A small part ($E = 120$ μJ) of the uncompressed radiation of a 1 kHz Ti:sapphire laser (780 nm, 8 ps) created a plasma on the C_{60}-containing target. The high-intensity 3.5 fs pulses were focused into the plasma plume to generate high-order harmonics. The delay between plasma initiation and the arrival at the plume of the few-cycle pulse was fixed at 34 ns. The HHG radiation was analyzed by an XUV spectrometer. We investigated two types of targets: (i) C_{60} powder, which was glued onto a glass substrate or onto a rotating aluminum rod, and (ii) bulk graphite for plasma harmonic generation. The harmonic spectra from plasmas produced on the bulk graphite and fullerene powder glued onto the glass surface are presented in Fig. 7.1 in the case of laser wavelength 780 nm. Harmonics up to the 29th order were obtained from the fullerene plasma. Note that the harmonic efficiency in the case of the graphite plasma was about five times higher than with the case of fullerene plasmas, most likely due to the higher concentration of emitters in the former case, as has also been reported in [3].

Here, we also present the results of comparative studies of the HHG in fullerene plasma using few-cycle (3.5 fs) and multi-cycle

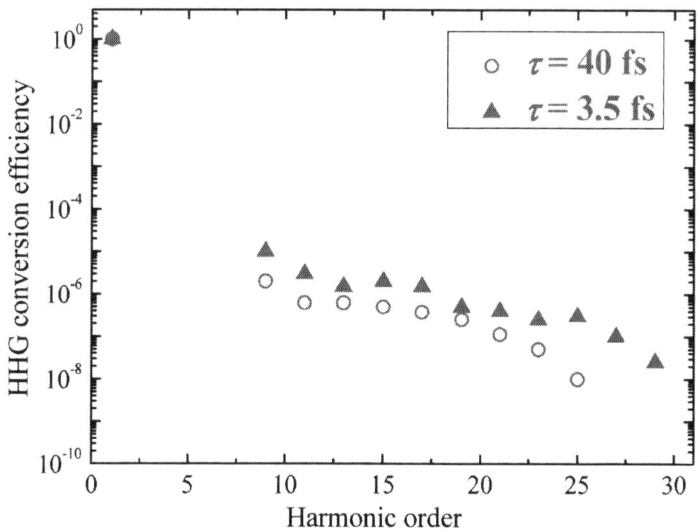

Figure 7.2 Comparison of HHG conversion efficiencies from fullerene plasma using the 3.5 fs (triangles) and 40 fs (circles) probe pulses. Here the laser wavelength is 780 nm. Reproduced from Ref. 2 with permission from Optical Society of America.

(40 fs) pulses (Fig. 7.2) of 780 nm radiation. In the latter case, we used a 1 kHz repetition rate Ti:sapphire laser generating 4 mJ, 40 fs pulses, which were used for the HHG in various laser plasmas. The HHG conversion efficiency for the 40 fs pulses at the beginning of plateau range in the case of the plasma plume containing fullerenes was estimated to be nearly 5×10^{-6} using a comparison with the HHG conversion efficiency in a silver plasma, which has previously been reported at similar experimental conditions to be 1×10^{-5} [4]. One can note that the cutoff in the case of longer pulses (25th harmonic) was shorter with regard to the few-cycle pulses at similar intensities of these pulses in the plasma plume.

The problem with using a fullerene powder-containing target is the shot-to-shot instability and rapid decrease of the harmonic yield, due to the abrupt change in the target morphology following ablation. Such instabilities limit applications of this radiation source and greatly hamper efforts to measure the pulse duration of the harmonic emission, which typically requires a large number of laser shots. A solution to this problem might be the use of long homogeneous tapes containing fullerenes, which continuously moves from shot to shot to provide a fresh surface for each next laser pulse. The use of rotating targets containing C_{60} is another method for improvement of the harmonic stability [5] from this medium, which was implemented in this work. We used 15 mm diameter aluminum rods as the substrates onto which the fullerene powder was glued. These rotating targets considerably improved the stability of harmonics from C_{60}-containing plasmas compared with fixed targets fabricated by gluing fullerene powder to a glass substrate. We found that once the target rotation is stopped, the harmonic efficiency from the fullerene plasma decreased to the noise level within 1–2 s.

This target fabrication method could be very useful in the case of powder-like targets (fullerenes, metal nanoparticles, organic powder-like samples, and so on). Previously, even at 10 Hz pulse repetition rate, the ablation of powdered targets that were not moved led to their rapid degradation and abrupt decrease of harmonic yield [4]. With this new technique, one can glue the powdered material onto the rotating rod to improve the stability of the harmonic emission. At the same time, one can note that the stability of harmonics from such powder-like targets is still worse than from rotating bulk metal rods due to the faster degradation of the target surface in the former case.

We did not estimate the density of fullerenes on the rotating target. In principle, it should be the same as the density of powder, as we just glued the fullerene powder without additional pressing. The thickness of powder layer was 2 mm. The uniformity was maintained with the accuracy better that 0.1 mm.

Due to small concentration of fullerenes in the plasma plume, the amount of ablated material was insignificant even at 1 kHz pulse repetition rate. So, the main difference between the rotating and non-rotating targets was the thermal conditions of heating spot. In the case of fixed target, the melting bath appeared after 1000 shots (e.g., 1 s of ablation), which considerably worsened the process of plasma formation. Once the target started to move, the previously heated area cooled down and again could be used for efficient ablation and harmonic generation. We observed this phenomenon from many targets. So, here we were able to improve the stability of harmonics (and correspondingly stability of plasma formation conditions) not just by using the fresh (non-ablated) surface but by changing the conditions of overheating of the same spot of the target. On the basis of our experiments with different (rotating and fixed) targets, we concluded that the overheating of the same spot worsens the conditions for stable plasma formation. The morphology of the heating spot did not remain intact during these experiments. At the same time, the creation of craters in the case of rotating target occurred after considerably longer period than with static targets.

Conversion efficiency studies in these two plasmas showed advantages of HHG in the case of graphite plasmas compared with fullerene plasmas [3]. This can be explained by the higher particle density in the graphite plasma. The concentration of fullerenes is below 10^{17} cm^{-3} [6], while our density estimates for carbon plasmas based on a three-dimensional molecular dynamical simulation of laser ablation of graphite using the molecular dynamics code ITAP IMD [7] showed that for heating by 8 ps laser pulses, the graphite plasma density can reach 2.6×10^{17} cm^{-3} at the moderate ablation intensity of 2×10^{10} W/cm^2. Another reason for the observed superior features of graphite plasma harmonics could be the production of clusters during laser ablation, though their involvement in HHG requires additional studies, including time-of-flight measurements.

In order to analyze the harmonic yield from the fullerene plasma using the midinfrared laser source, we used an optical parametric oscillator pumped by the 40 fs Ti:sapphire laser. The OPA provided signal pulses with duration of 35 fs in the 1200–1600 nm spectral range with a maximum energy of 1.7 mJ at nearly 1300 nm. Figure 7.3 shows the comparison of fullerene harmonic spectra generated in the case of 1300 and 780 nm multi-cycle probe pulses. Harmonics up to the 41st order (Fig. 7.3, bottom panel) were observed in the case of 1300 nm probe pulses at the conditions of optimal plasma formation using the heating 160 ps pulses. The application of less intense, longer wavelength (1400 nm) pulses available by tuning the OPA did not result in an extension of harmonic cut-off compared with the case of probe 1300 nm pulses. This observation suggests that the

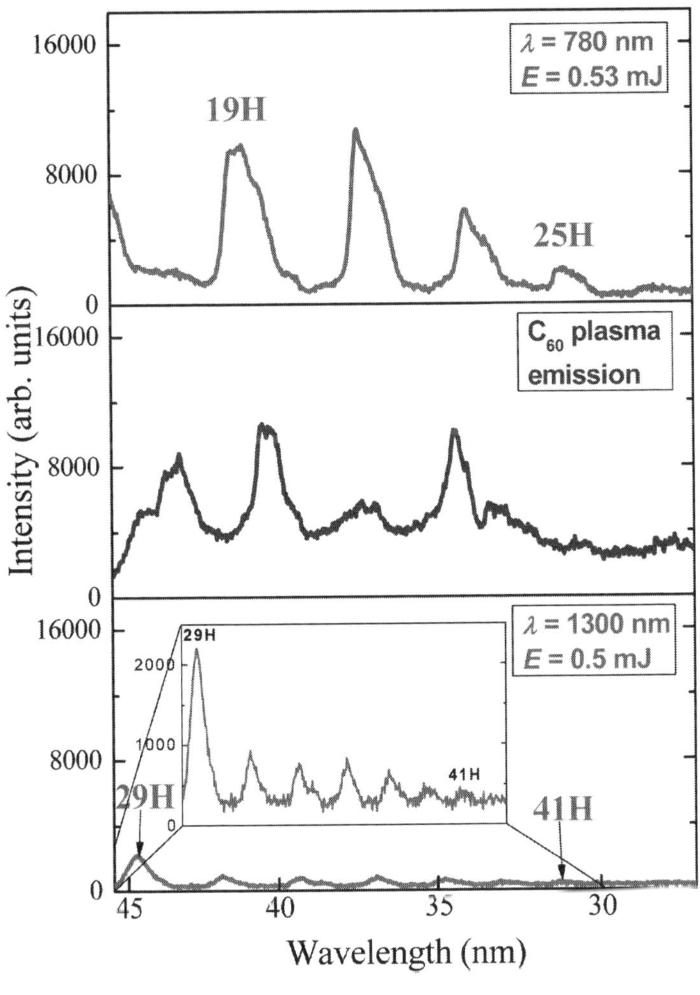

Figure 7.3 Comparative harmonic spectra from fullerene-contained plume using the 780 nm (upper panel) and 1300 nm (bottom panel) multi-cycle pulses (intensity of heated 160 ps pulses $I_{ps} = 1\times 10^{10}$ W cm^{-2}). The intensity axes are on the same scale allowing a direct comparison between the three cases. The middle panel shows C_{60} plasma emission spectrum at over-excitation of target by 20 ps pulses ($I_{ps} = 4\times 10^{10}$ W cm^{-2}), without further excitation by femtosecond probe pulses. Reproduced from Ref. 2 with permission from Optical Society of America.

harmonic generation occurred under saturated conditions, with the expectation of stronger harmonics once the micro and macro-processes governing frequency conversion are optimized. Over-excitation of target by 160 ps pulses (Fig. 7.3, middle panel) led to appearance of plasma emission in the 25–45 eV range of photon energies ($\lambda = 27$–50 nm). At these conditions, no harmonics were observed during propagation of femtosecond pulses through such over-excited plasma.

The harmonic spectrum up to the 25th order in the case of 780 nm, 40 fs probe pulses is presented in the Fig. 7.3 (upper panel). One can clearly see the extension of harmonic cutoff (from the point of view of highest harmonic order) in the case of longer-wavelength driving pulses by comparing the HHG spectra using the 780 and 1300 nm pulses, while the extension of cutoff energy was insignificant.

We maintained approximately equal energies of driving pulses in these cases (0.53 mJ for 780 nm pulses and 0.5 mJ for 1300 nm pulses). The intensity of 780 nm radiation in the plasma area was calculated to be $I_{fs} = 4 \times 10^{14}$ W/cm^2. The pulse durations of these sources (780 and 1300 nm) were approximately same (40 and 35 fs, respectively). Probably due to phase modulation and propagation through multiple optical elements in optical parametric oscillator-amplifier, the diameter of 1300 nm beam in the plasma plume was bigger than in the case of 780 nm beam. The corresponding lower intensity of 1300 nm radiation could be responsible for less expected extension of harmonic cut-off energy. From the cutoff formula, one could expect the generation of harmonics up to the 47th order (in the case of 780 nm radiation), well above the observed cutoff (25th harmonic), which probably points out the difference in expected and actual intensity in the plasma area at the optimal conditions of HHG. The reason in this discrepancy could be related with the self-defocusing of driving pulses in the medium containing free electrons. Below, we also discuss on the cutoff law and on its limit of validity at our experimental conditions (see also Section 2.2).

In atoms, the maximum emitted photon energy (giving the cutoff position) is described by the well known relation $E_M = I + 3.17U_p$ with I the ionization energy and U_p the ponderomotive energy. By invoking the three step model [8], this formula can be derived (by use of mere energy conservation and Newton laws) under the assumptions that the laser field amplitude is constant. Rapidly varying laser pulses can considerably shift the cutoff [9, 10]. We also know that molecules support more returning trajectories than atoms and allow the existence of several plateau [11]; it is not therefore easy to disentangle the effects of pulse shape and presence of a molecule and to state a general law for the position of the cutoff; the $E_M = I + 3.17U_p$ law can therefore be used only as a touchstone.

We observed that the plasma harmonic yields from the 780 and MIR probe pulses are consistent with the predicted single-atom harmonic intensity wavelength scaling $I_h \propto \lambda^{-5}$, which arises due to electron wavepacket spreading before recollision. The harmonic efficiency of the XUV radiation in the range of 30–50 nm driven by MIR pulses was seven to 15 times less compared with the case of 780 nm probe pulses, which is comparable with the expected ratio between harmonic intensities from these sources $(1300/780)^5 \approx 12.7$, followed from above rule, assuming approximately equal energies of the 780 and 1300 nm pulses (0.53 and 0.5 mJ, respectively).

Previous HHG studies in fullerenes were performed using the multi-cycle pulses (30 fs, 48 fs, and 110 fs [4]). The stability of C_{60} against fragmentation in multi-cycle laser fields leads to fast diffusion of the excitation energy. Even better conditions can occur in the case of few-cycle pulses used for fullerene HHG. In that case, fullerenes can withstand the influence of the strong field of few-cycle pulses due to the increase of the ionization saturation intensity as the pulse duration is decreased. This can further increase the diffusion of the excitation energy within the fullerenes due to their very large number of internal degrees of freedom. The increase of energy diffusion is expected to lead to a decrease in the disintegration of fullerenes, which enhances the probability of harmonic emission from these molecules.

The graphite plasmas, which were compared in our experiments with C_{60}, can easily be aggregated during laser ablation, thus leading to nanoparticles in the laser plume. In that case, the comparison of two clustered species (large 5–20 nm carbon clusters and 0.7 nm C_{60}) can lead to their different nonlinear optical response once the interacting laser pulse becomes compressed from the multi-cycle to few-cycle duration. Our results show that for the shortest laser pulses, the HHG cutoff for fullerene is extended, while the harmonic intensity is reduced compared with "carbon harmonics." The intensity of the probe femtosecond pulse is an important parameter for optimizing the HHG from C_{60}. Increasing the intensity of the probe pulse did not lead to an extension of the harmonic cutoff from the fullerene plume, which is a signature of HHG saturation in the medium. Moreover, at relatively high probe intensities, we observed a decrease in the harmonic output, which can be attributed to the phase mismatch caused by propagation effects.

The stability of C_{60} molecules against ionization and fragmentation is of particular importance, especially for their application as a medium for HHG using laser pulses of different duration. The fullerene molecules ablated off the surface should be intact when the probe pulse arrives. Hence, the heating pulse intensity also becomes a sensitive parameter. At lower intensities, the concentration of the clusters in the plume would be low, whereas at higher intensities, one can expect fragmentation. This phenomenon is observed when the heating pulse intensity on the surface of fullerene-rich targets is increased above the critical value (Fig. 7.3, middle panel). The abrupt reduction in harmonic intensity in that case can be attributed to phenomena such as fragmentation of fullerenes, an increase in free electron density, and self-defocusing of the probe laser, all of which are expected to reduce the efficiency of HHG.

7.2 Single Active Electron Simulation of Harmonic Generation in C_{60}

Fullerenes are now considered very important targets in plasma ablation HHG experiments. C_{60} fullerene is the most widely used fullerene for HHG [6]. The main attractive feature of C_{60} is a broad plasmon resonance in VUV domain. The corresponding HHG spectra can have a group of enhanced harmonics in the range of this resonance (20 eV) with conversion efficiency above 10^{-5}.

However, further extension and optimization of fullerene HHG experiments require good theoretical background. From theoretical point of view, fullerenes are not only large molecules but also the smallest clusters with metallic properties, which can still be simulated directly in time-dependent phenomena, such as HHG. C_{60} is known to have 240 valence electrons, 200 of them are localized σ-electrons and remaining 40 are delocalized π-electrons, but in a numerically exact calculation, they are all important. This large number of electrons makes any direct numerical time-dependent Schrödinger equations (TDSE) solutions impossible. The time-dependent density functional theory (TDDFT) [12] studies of C_{60} are much more efficient; however, pseudopotential-based simulations of C_{60} in intense laser field can easily overestimate ionization because the highest occupied Kohn-Sham orbitals are too close to continuum [13]. Up to now, success in C_{60} TDDFT investigation has been achieved mostly by means of jelly-sphere model [14, 15]. It has been found in [16] that with a few exceptions, the eigenfunctions of the valence band and the bottom of conduction band are characterized mainly by a single angular momentum quantum number l. Jelly potential is rather simple and, being semi-empirical, it can describe time-dependent properties very well, but it is still rather computationally demanding because even in TDDFT approach, simulation of 240 electrons is approximately 240 times more demanding than SAE simulation on the same numerical grid representation.

Recently, a new SAE paradigm of fullerene has been suggested [17] and applied for HHG simulation for various pulses with results, which agree qualitatively with experiments. It was based on the fact that, in real experiments, C_{60} molecule can absorb up to 50 eV before being ionized. The fullerene was represented as a SAE confined in general on a two-dimensional surface with radius R = 6.7 atomic units (a.u.). The main advantage of this model is that corresponding TDSE can be solved numerically very easily. Although it has been shown in [18] that nonlinear response of a system where no ionization is possible can have properties of HHG response, it cannot be explained by simple-man model [8] and the physical meaning of a completely confined electron is rather unclear. Also, in HHG experiments, C_{60} is ionized during interaction with strong femtosecond pulse. Later, similar approach has been used to simulate HHG from hydrogen atom very exactly and efficiently

[19]. In addition to decomposition of electron angular momentum in series of orbital quantum number l, a radial degree of freedom in an unsmoothed Coulomb potential has been added.

This motivated us to extend the rigid sphere fullerene paradigm to well-understandable three-dimensional systems that can actually be simplified to two-dimensional systems in case of linearly polarized pulses. Our model differs from [17] by inclusion of radial groundstate and movement of electron. It differs from [19] as well because electron's groundstate is approximately on the surface of rigid sphere with radius $R_0 = 6.7$ a.u. and electron cannot go inside this sphere. We also try to solve the TDSE equations without decompositions of potential for angular degree of freedom. We should note that our simplified model is not an exact fullerene, but rather a fullerenelike system, but the main features of fullerene can be reproduced correctly.

The choice of radial potential is the key feature of our model. Fullerene molecule is actually rather sparse object with average C–C distance of 1.44 Angstrom. Thus near the surface of fullerene an electron is very close to its parent atom and we can safely try to use softened Coulomb potential with minimum at $R_0 = 6.7$ a.u. so that one-dimensional ionization energy is close to 7.6 eV. At $R_1 = 7$ a.u. the potential was changed by unsmoothed non-shifted Coulomb potential of the sphere, because the positive charge can be considered as point charge in the centre of the sphere. The difference between the attraction of one atom and the whole molecule explains the change in potential at $R_1 = 7$ a.u.

The angular degree of freedom is the same as in [17]. We study only interactions with linearly polarized pulses, so our system can be reduced from spherical to polar in this special case. For the rest of the article atomic units will be used unless other explicitly stated. The resulting general form of TDSE for the considered fullerenelike system is:

$$i\frac{\partial\Psi(r,\theta,t)}{\partial t} = -\left(\frac{\partial^2}{\partial r^2} + \frac{L^2}{2r^2} - V(r) + r\cos\theta E(t)\right)\Psi(r,\theta,t)$$

$$V(r) = \begin{cases} -\dfrac{1}{\sqrt{(r-6.7)^2+0.05}}, & 6.7a.u. \le r \le 7a.u. \\ -\dfrac{1}{r}, & r > 7a.u. \end{cases} \tag{7.1}$$

We use discrete variable representation of the wavefunction [20], so the radial degree of freedom is represented by particle-in-a-box eigenfunctions

$$\begin{cases} \varphi_1(x, t) = A(t) \sum_{i=1}^{N} \sqrt{2/L} \sin(j\pi(x - x0)/L) \; 0 \, for \; x_0 \leq x \leq x_{N+1} \\ 0 \qquad\qquad\qquad\qquad else \end{cases}, \tag{7.2}$$

and the angular degree of freedom is represented by Legendre-type polynomials:

$$\varphi_2(\theta, t) = B(t) \sum_{l=1}^{M} \sqrt{\frac{2l+1}{2}} P_l^0[\cos(\theta)] \tag{7.3}$$

Our representation of wavefunction results in the decoupling of angular and radial degrees of freedom, which is not a simplification of the system, because the TDSE solution is basis-independent when basis for the wavefunction is sufficiently large, as in this case. The equations of motion are thus the TDHF equations:

$$\begin{aligned} &\Psi(r, \theta, t) = a(t)\varphi_1(r, t)\varphi_2(\theta, t) \\ &\langle \varphi_1\varphi_2 | \, i\dot{a}\varphi_1\varphi_2 + ia\dot{\varphi}_1\varphi_2 + ia\varphi_1\dot{\varphi}_2 - Ha\varphi_1\varphi_2 \rangle = 0 \\ &\langle \varphi_1 \mid \dot{\varphi}_1 \rangle = \langle \varphi_2 \mid \dot{\varphi}_2 \rangle = 0 \end{aligned} \tag{7.4}$$

Complicated effects, such as multiple molecular orbital effect, multiple center interference, electron-nuclear correlation are not supported by this simplified model. However, our molecule is close to spherically symmetric one so molecular orbital effect and multiple center interference can be averaged without loss of generality. Although electron-electron correlation can be very important for ab initio simulation of resonant HHG effects, electron-nuclear correlation in experiments is too weak because electron interacts with the whole 60-atom molecule.

HHG is actually only the work of laser field for de-acceleration of an electron. So it is the acceleration form that should be computed and Fourier-transformed to get the HHG spectrum as exactly as possible. It has been shown in [19] that velocity and dipole forms can have very significant, but fully predictable deviations from exact results of HHG depending only on the laser pulse parameters. So the expected quantity is:

$$\langle \Psi(r, \theta, t) | - \frac{\partial H}{\partial (r\cos\theta)} | \Psi(r, \theta, t) \rangle \tag{7.5}$$

and its exact Fourier transform gives us the power spectrum. However, discrete Fourier transforms cannot correctly distinguish any spectral component with amplitude less than 0.001 of the dominant spectral component, even in a simple mixture of two perfect sine pulses, so for prediction of harmonics with intensity less than 10^{-6} of the main pulse we use piecewise least-squares approximation, which expands the expectation value of dipole acceleration at every short time domain in a sum of pure sinusoids. Although information about the phase of harmonics is lost in this process, the intensities are obtained mathematically exactly and we get the best correspondence with the experiment, which can ever be achieved.

We use the capabilities for direct solution of TDSE of freely available on request Heidelberg multiconfigurational time-dependent Hartree (MCTDH) package [20–22] to solve this two-dimensional equation exactly. We do not need to reduce the problem to one-dimensional case by expansion of angular momentum operator in infinite series of l, because the possible computational gain is not worth rewriting the potential for every additional l. However, for larger systems, this decomposition, introduced in [17] can be useful. MCTDH package is capable of solving much more complicated problems than simple 2-dimensional case, and actually we do not go beyond simple time-dependent Hartree for two degrees of freedom in this article, but it's a very reliable program and this simple problem is solved very efficiently.

We performed HHG calculations for pulses, which are close to those used in previously reported HHG experiments in fullerene-contained plasmas. Experimental conversion efficiency at the plateau range was estimated as close to 10^{-4} for the most enhanced harmonic in all experiments. First calculations were performed for the 150 fs pulse with central wavelength 792 nm and averaged intensity 7×10^{14} W cm^{-2}. The resulting HHG spectrum is given in Fig. 7.4, together with reported data for this pulse [23]. One can see that in our studies only 11–15 harmonics have been sufficiently enhanced, while in experiment harmonics up to 27th were rather strong. This can be attributed to differences in spectral properties of our model system and real fullerene.

Next calculation was performed for the 48 fs pulses with central wavelength 800 nm and averaged intensity 7×10^{14} W cm^{-2}. The

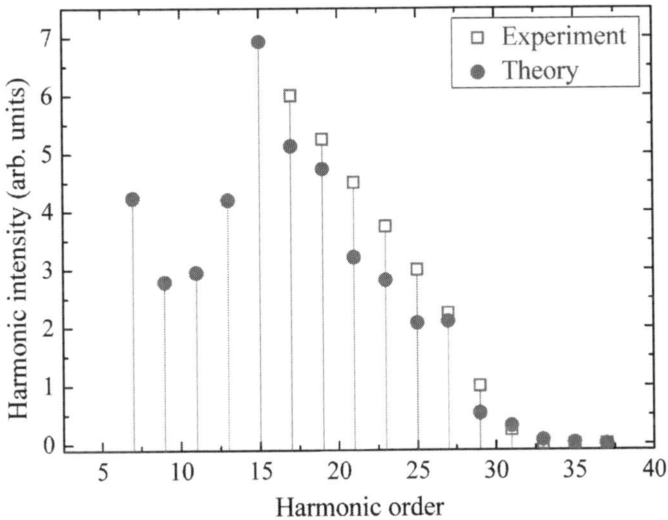

Figure 7.4 Calculations (circles) of fullerene HHG using the 150 fs pulse with central wavelength 792 nm. These results were compared with experimental data (squares).

resulting HHG spectrum is given in Fig. 7.5; the corresponding reported experimental results [24] are shown in this figure. Again, the qualitative coincidence in the harmonic intensity distribution is relatively good, except for the 15th harmonic, which is not so enhanced in our simulations.

In the last calculation we used the pulse with duration 35 fs and wavelength 793 nm and averaged intensity 7×10^{14} W cm^{-2}. The results from our calculations (Fig. 7.6) have more reliable correspondence to the experimental results [25] than the results of TDDFT simulation from that work. This difference with TDDFT results can be explained by the fact that in those TDDFT simulations monochromatic pulses which are unreachable in experiments were used and propagation effects were taken from a rough external estimate. Thus, correct treatment of the pulse is much more important for HHG simulations than exactness of numerical approach.

One can clearly see a good qualitative correspondence between the theoretical and experimental spectra in intensity of harmonics.

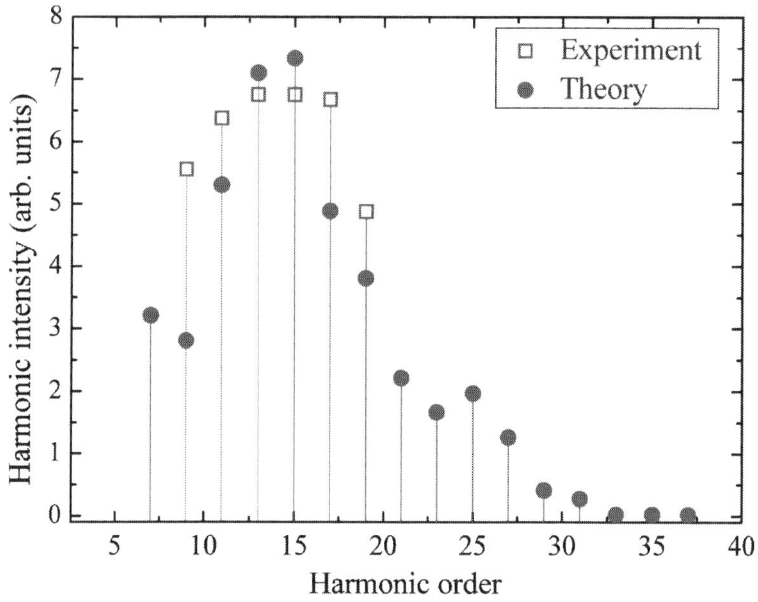

Figure 7.5 Calculations (filled circles) of fullerene HHG using the 48 fs pulse with central wavelength 800 nm. These results were compared with experimental data (squares).

It has been shown in [25] that propagation effects do not change the HHG output qualitatively. As a result our simplified model can give reliable estimates of C_{60} HHG experiments with a low computational effort, which allows in future solving time-dependent Schrödinger equation simultaneously with propagation equation to get the most correct description of plasma HHG experiments. This computational simplicity is also for numerical optimization of single-molecule HHG yield by means of optimal control theory simulations, where a lot of propagation runs should be performed. However, our model should be used with care for tasks other than HHG simulation, because even the quantitative correspondence in one kind of processes cannot prove that the model is valid for all possible phenomena.

The most important feature of any model of C_{60} fullerene is its ability of reproducing a broad plasmon absorption band at 60 nm with FWHM of 10 nm. The spectrum of eigenfunctions is actually not the system's spectrum, because the stability of eigenstate should

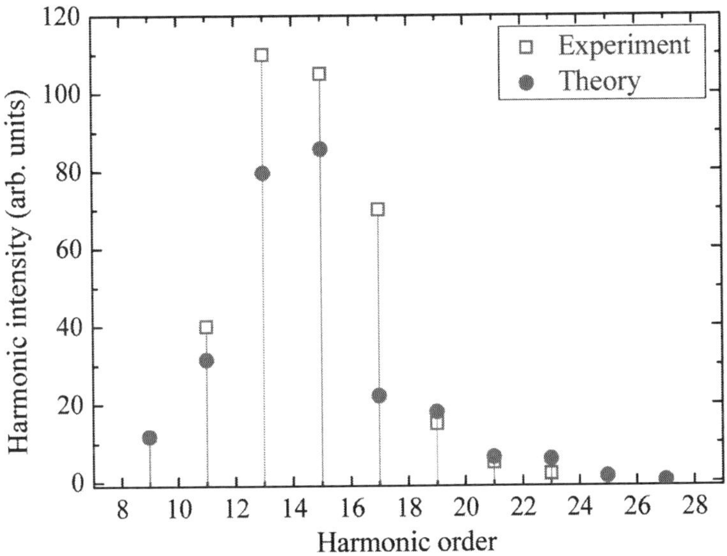

Figure 7.6 Calculations (filled circles) of fullerene HHG using the 30 fs pulse with central wavelength 793 nm. These results were compared with experimental data (squares).

be conserved in time. To obtain a reliable absorption spectrum of the system we performed propagation of initially perturbed wavefunction in the absence of any external field. The spectrum was taken in the same way as the HHG spectrum, with the only difference that the function $\langle\Psi(r, \theta, t_0)| r \cos\theta |\Psi(r, \theta, t)\rangle$ was approximated to a sum of Gaussian-broadened sinusoids. The resulting spectrum (Fig. 7.7) indeed shows enhanced absorption in regions of 9–17th harmonics of Ti:sapphire lasers (14–26.5 eV) so it can be used to qualitatively describe HHG from fullerenes. Further increase of simulation time did not change the positions of absorption peaks of the spectrum and their width will be constant anyway due to chosen way of spectral estimate with a piecewise least-squares approximation to equally broadened signals, so the spectrum is well-converged. It has been pointed out in [26] that effect of plasmon resonances can enhance the intensity of the pulse near the particle, so the effect of local field is probably the mechanism of HHG enhancement near plasmon resonance in our model as well.

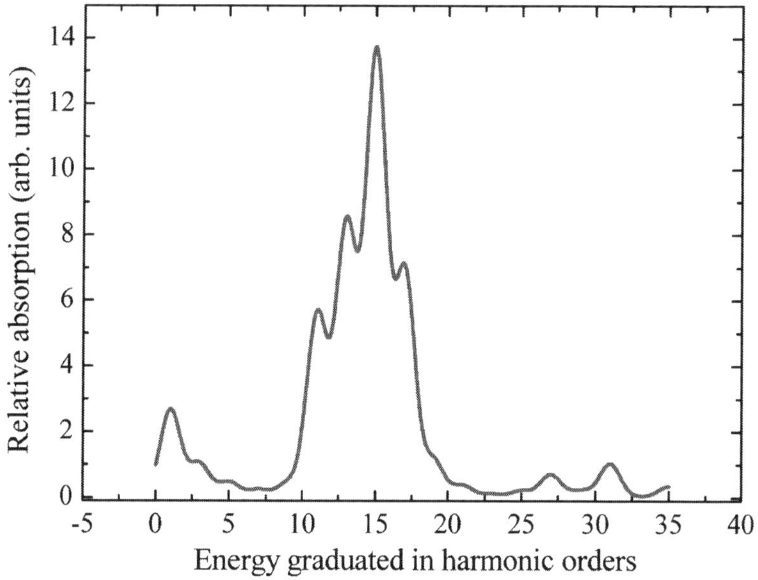

Figure 7.7 Calculations of fullerene absorption. The x-axis is presented in the units of harmonics from Ti:sapphire lasers.

7.3 Ablation of Nanoparticles and Efficient Harmonic Generation Using 1 kHz Laser

Methods for improvement of low efficiency of the conversion of coherent photons from commonly used laser sources towards the short wavelength range include both those related with monitoring of the microprocesses and those related with modification of macroprocesses occurring during laser-matter interaction. In the past, HHG, as a most commonly used process for conversion of the frequency of laser radiation from the infrared range to extreme ultraviolet range, has demonstrated various ways for improvement of harmonic conversion efficiency. Most frequently used gaseous media and surfaces, as well as recently emerged plasma media, are the subjects of investigation of these methods. Initial conversion efficiencies (η) in these media did not exceed 10^{-8}–10^{-7}. These values of converted photons were absolutely insufficient for any scientific and practical applications of those XUV sources. All the history of gas HHG (starting from 1987 [27–29]) and plasma HHG (starting from 1992 [30]) is related with a search of advanced methods for the improvement of harmonic conversion efficiency. Nowadays, this search of new methods still remains one of the main goals of nonlinear optics.

The very low values of η in HHG are related with few peculiarities of conversion technique in isotropic low-dense media. Conversion efficiency of laser radiation in the condensed media (so-called nonlinear crystals) has started with the same low efficiencies [31] and soon after that was considerably improved thanks to the development of the technique of creation of the optimal phase conditions between the converted and converting waves. This macroprocess is related with appropriate matching of the wave vectors of two electromagnetic waves, when the ensembles of photons coherently converted in second, third, fourth, and fifth harmonics in the visible and near ultraviolet ranges. The application of large amount of emitters comprising the whole condensed matter at the particles density of 10^{23}–10^{24} cm^{-3} allowed achieving extremely high conversion efficiencies (in the range of 10^{-1}–6×10^{-1}). Most of these achievements were due to the high concentration of emitters and application of propagation (i.e., macroscopic) effects.

Movement towards the sorter wavelength region did not allow the application of condensed matter due to considerable absorption of converting photons inside this medium. Initially gaseous and then, starting from recent time, plasma media replaced the condensed materials as the sources of harmonics. Very low concentrations of emitters in these cases (~10^{17}–10^{18} cm^{-3}) led to drastic decrease of the η of HHG due to quadratic dependence of this process on the density of converting media.

Starting from this level of conversion efficiencies (10^{-8}–10^{-7}), various approaches were proposed and applied for enhancement of the HHG efficiency. Most of them were related with the improvement of the influence of propagation effects on the process of wavelength conversion. Those include the optimization of Gouy phase, addition of the media with positive or negative dispersion for achieving phase matching conditions, application of quasi-phase matching using few gas jets, aperturing of laser beam, etc. ([28, 32–36], also see monograph [37] and references therein).

The microprocesses, for example, the processes related with the response of single atoms/ions but not ensembles, also were treated for these purposes. Most important of them was the involvement of resonance processes in the enhancement of the nonlinear optical response of the medium. Some achievements of using resonances were demonstrated at the early stages of low-order harmonics generation in the vacuum ultraviolet range (80–100 nm [37]). However, the use of limited number of gaseous media did not allow the application of this effect in the XUV range. In the meantime, plasma media were exceptionally suitable for application of this approach for the enhancement of single harmonic in the short-wavelength range. While most successful attempts for improvement of η using various above-mentioned approaches related with macro-processes led to the enhancement of HHG conversion efficiency at the conditions of harmonic generation far from resonances up to the levels of 10^{-5} both in gases [36] and plasmas [38], the resonance-related improvement of this process allowed achieving the 10^{-4} conversion efficiency [39] for some specific cases, in particular in indium plasma. Further improvements of HHG efficiency in these media are under consideration.

The way to overpass the notoriously low amount of particles participating in HHG is the application of clusters. The increase of concentration of emitters can drastically improve the η of harmonics. First attempts on this direction were made using the gaseous clusters [40–45]. The improvements of η were related both with the increase of the concentration of harmonic emitters and specific properties of clusters. Those included, in particular, the increase of cross-section of the recombination of accelerated electron with parent particle in the case of clusters compared with atoms. Another, still unexplored and fully understood, idea was the application of local field enhancement in the vicinity of clustered media and use of plasmonic properties of nanoparticles and nanoantennas in gaseous media [46–49]. In the meantime, the application of surface plasmon resonance for enhancement of harmonics was realized in the plasma HHG studies of fullerene media [25, 50]. Further, the ablation method for creation of the plasma media containing large amount of nanoparticles became available due to appearance of various nanoparticles in the market.

Applications of ablated nanoparticles for enhancement of HHG were confirmed in many laboratories ([51–53], see also the review of those studies [54]). The reported improvement in η using various metal clusters showed the attractiveness of this approach.

These plasma HHG studies of clustered media were carried out using relatively low (10 Hz) pulse repetition rate lasers. The problem of maintaining the stable plasma plume containing nanoparticles was a main obstacle of this technique. In the meantime, the increase of average power of harmonics could be further achieved with the use of high pulse repetition rate lasers. In these conditions, a decrease of nanoparticles concentration during multiple ablation of cluster-contained targets can considerably worsen the HHG. A search of the methods allowing to overcome this obstacle is an important goal for further improvement of the η of HHG and average power of harmonics. It will also allow the analysis of various, particularly plasmonic, properties of clusters as the sources of harmonics.

In this section, we show a method for the stabilization of laser plasma plumes containing metal nanoparticles, analyze the morphology of ablated media, and show the advantages of cluster plasma HHG at high pulse repetition rate [55]. Particularly, we report the first studies of the dynamics of nanoparticles ablation and HHG from this ablation at 1 kHz pulse repetition rate. Analysis of the morphology of the ejected aggregates has allowed to define the conditions that lead to the generation of a plasma which contains nanoparticles of the same characteristics as the ones used as target material. Comparison with harmonics generated in the ablation plasmas of bulk targets of the same metallic element, clearly demonstrates the superior efficiency of the nanoparticle-based targets. We discuss the mechanisms of efficient harmonic generation in cluster-containing media.

7.3.1 *Experimental*

A 1 kHz Ti:sapphire laser was used as the main laser driving source. A part of the uncompressed radiation was split off from the beam line prior to the compressor stage (780 nm, 1.3 mJ, 160 ps) and used for ablation of the targets placed in a vacuum chamber 500

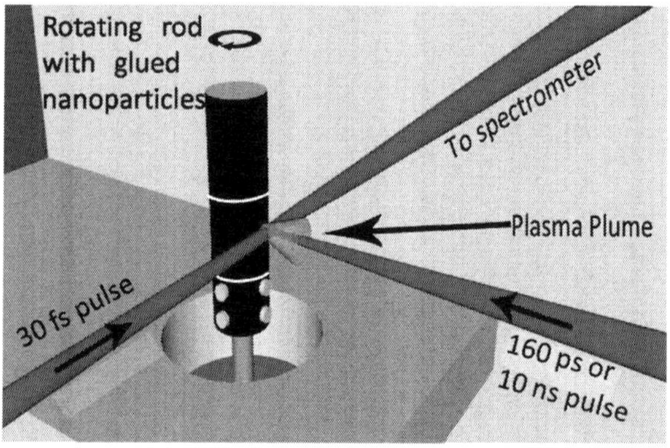

Figure 7.8 Experimental setup for ablation of the rotating aluminum rod covered by a layer of different metallic nanoparticles. The nanoparticles were glued on the surface of the rod limited by the two curved white lines. Reproduced with permission from Ref. 55. Copyright 2013 American Physical Society.

μm (Fig. 7.8). After a delay of ~40 ns, required for formation and expansion of the plasma plume away from the surface of the target, the compressed laser pulse (780 nm, 1 mJ, 30 fs) was focused onto the plasma to generate the high-order harmonics.

Small-sized (20–25 nm) Al, Ag, and Cu nanoparticles were used in most of these experiments. We also used large (100–150 nm) silver and aluminum nanoparticles. We aimed at comparing the characteristic morphology and size of the initial nanoparticles constituting the ablation target material with those of the nanodeposits collected on silicon wafers placed at a distance of 4 cm. The range of sizes provided also the case for studying the dependence with size of the nonlinear optical response of the corresponding laser plasmas. These powders were prepared as the targets by gluing on the aluminum rotating rods. The rotation speed of rods was 15 rpm. Rotation of the rod during ablation prevented overheating and damage of the same spot of the surface from repeated laser shots, thus ensuring more stable ablation conditions. This technique, which was described in previous chapters, was recently introduced in the case of bulk targets and fullerene powder and showed to be advantageous as compared with static targets [5]. We compared HHG in nanoparticle based plasmas and in plasmas generated by

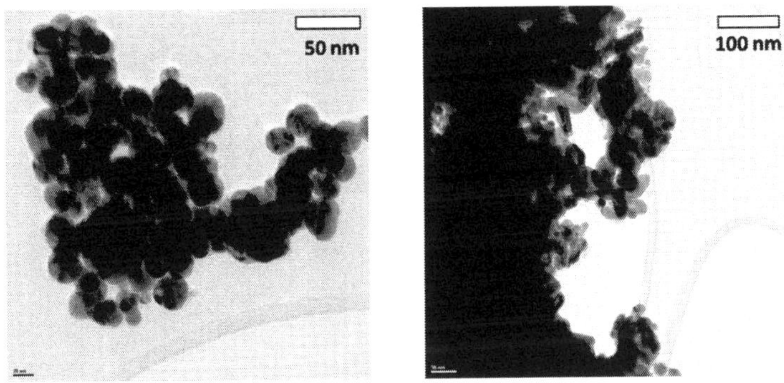

(a) (b)

Figure 7.9 TEM images of as-supplied (a) silver and (b) aluminum nanoparticles. Reproduced with permission from Ref. 55. Copyright 2013 American Physical Society.

laser ablation of bulk targets of the same metallic elements (i.e., Ag, Al, and Cu bulk rods). Characterization of the plasma debris collected on the silicon wafers placed 4 cm from the ablated target was carried out by scanning electron microscopy and transmission electron microscopy (TEM).

7.3.2 *Morphology of Ablating Nanoparticles and Plasma Debris*

The TEM images of the as-supplied silver and aluminum nanoparticles are shown in Fig. 7.9. The mean sizes of the corresponding individual nanoparticles are 25 and 20 nm respectively. In accordance with specification from manufacturer, these nanoparticles were prepared using the laser ablation technique in liquids. Notice that the original nanoparticles were combined in aggregates, containing thousands separate clusters (see Fig. 7.10 showing the SEM images of aggregates of Ag, Al, and Cu nanoparticles). These nanoparticle powders were glued using superglue on the aluminum rotating rod, with further polishing of the glued surfaces. We ablated the surface of the rod containing those nanoparticles using 160 ps pulses of Ti:sapphire laser, while in some cases we also ablated these surfaces

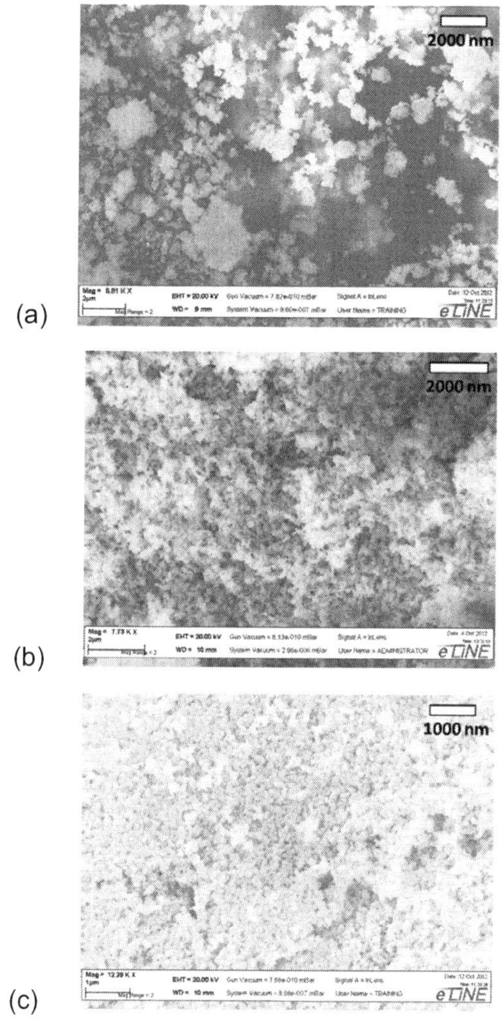

Figure 7.10 SEM images of (a) silver, (b) aluminum and (c) copper aggregates of nanoparticles before ablation. Reproduced with permission from Ref. 55. Copyright 2013 American Physical Society.

using the 10 ns pulses from Nd:YAG laser, to compare the variations of morphology and nonlinear optical response for different ablating conditions (160 fs pulses ablating at 1 kHz and 10 ns pulses ablating at 10 Hz pulse repetition rate). The debris from ablated areas were collected on the Si wafers placed 40 mm from the rotating target. Figure 7.11 shows the SEM pictures of the debris of ablated Ag, Al, and Cu nanoparticles using the pulses of different duration (160 ps and 10 ns). One can see that the characteristic sizes of the deposited nanostructured layer resemble those of the initial pristine nanoparticle target shown in Fig. 7.10. Notice that we used moderate laser ablation intensities (3×10^9 W cm^{-2} in the case of 10 ns pulses and 1×10^{10} W cm^{-2} in the case of 160 ps pulses) corresponding to conditions of optimal harmonic generation from these media. Over-excitation of nanoparticle-containing targets did not result in continuous nanostructured deposits but rather in the deposition of chaotically appearing aggregates (not shown). Also, at high ablation intensities, the HHG signals decreased considerably or completely disappeared after few thousands shots, even when using a rotating target.

We also analyzed the deposits generated upon laser ablation of the different bulk metallic (Cu, Ag, Al) at the ablation conditions that ensured optimum HHG conversion efficiency in the corresponding ablation plumes. The ablation of different bulk metal targets did not lead to appearance of nanoparticles in plasma debris at optimal intensities of heating pulses. This term refers to the conditions when maximum HHG conversion efficiency was obtained from these plasma plumes. Nanoparticles formation through the laser ablation of metals is a very elaborated technique (for example [56]. This method uses strong laser fields and don't care about the presence of free electrons appearing in large amount during ablation. However, the free electrons considerably worsen the conditions for efficient plasma HHG (as in the case of gas HHG). That is why the plasma formation from bulk targets for HHG requires moderate heating laser fields leading to the formation of the plasma plume containing low-excited, low charged monoparticles. We observed the nanoparticles formation in the debris at over-heating of bulk silver (i.e., at non-optimal excitation of target surface and inefficient HHG from the silver plumes). This observation was related with

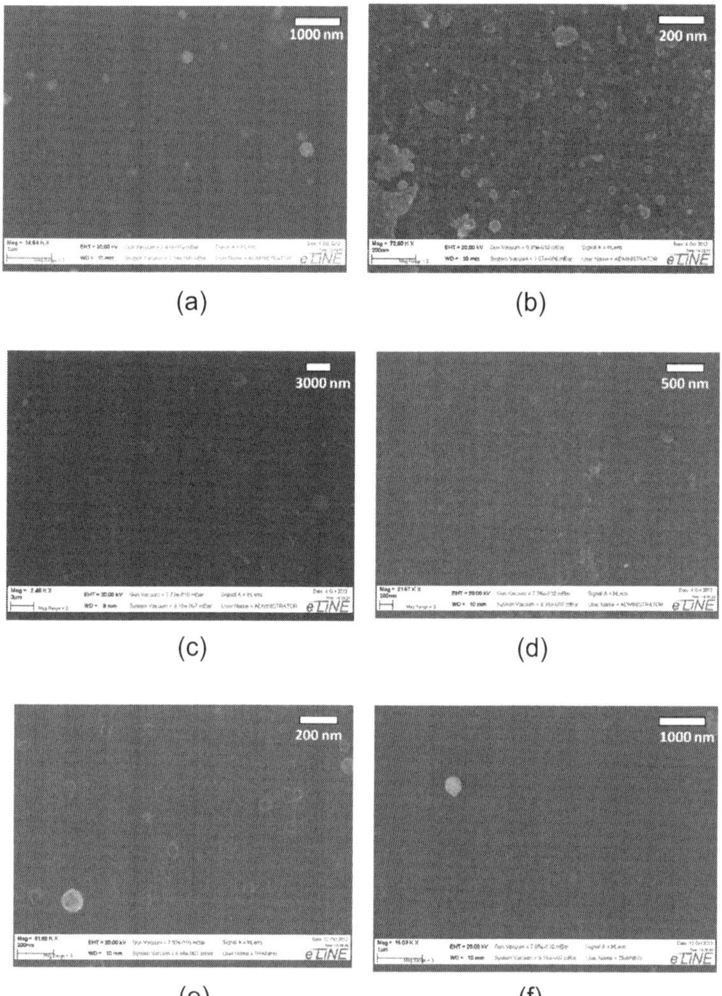

Figure 7.11 SEM images of material deposited on a Si substrate from ablation of targets containing the (a,b) Ag, (c,d) Al, and (e,f) Cu nanoparticles ablated with (a,c,e) 160 ps (1 kHz) and (b,d,f) 10 ns (10 Hz) pulses. The collection time for deposition was 3 min. Reproduced with permission from Ref. 55. Copyright 2013 American Physical Society.

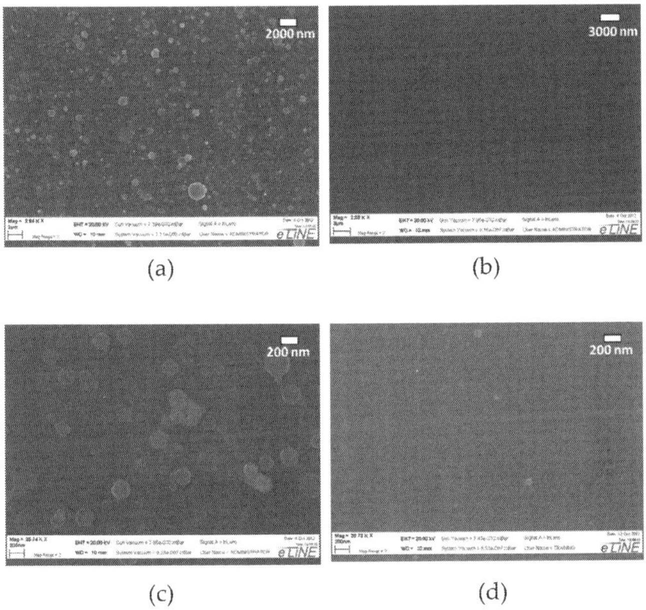

Figure 7.12 SEM images of material deposited on a Si substrate from ablation of targets constituted by bulk (a,b) silver, (c) aluminum, and (d) copper ablated with (a,c,d) 160 fs and (b) 10 ns pulses. Reproduced with permission from Ref. 55. Copyright 2013 American Physical Society.

ablation by 160 ps pulses (Fig. 7.12a), but not 10 ns pulses when we did not see nanoparticles formation even at strong fluences of laser radiation (Fig. 7.12b). The ablation of bulk Ag target by picosecond pulses led to appearance of broad variety of nanoparticles (from less than 10 nm to few hundred nm particles of spherical shape.

The increase of intensity of 160 ps heating pulse above some level of excitation of bulk targets ($>3 \times 10^{10}$ W cm^{-2}) led to considerable worsening of harmonic generation in all studied species. At these conditions, we observed the appearance of aluminum and copper nanoparticles in the debris deposited on the Si substrates (Figs. 7.12c and 7.12d), analogously to the case of silver target ablation. These observations demonstrate that the presence of nanoparticles in these plasma plumes did not compensate for the deteriorated conditions of over-ionized plasma containing large

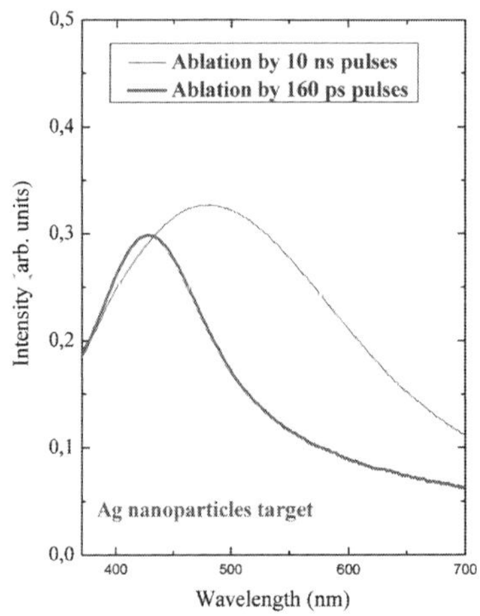

(a)

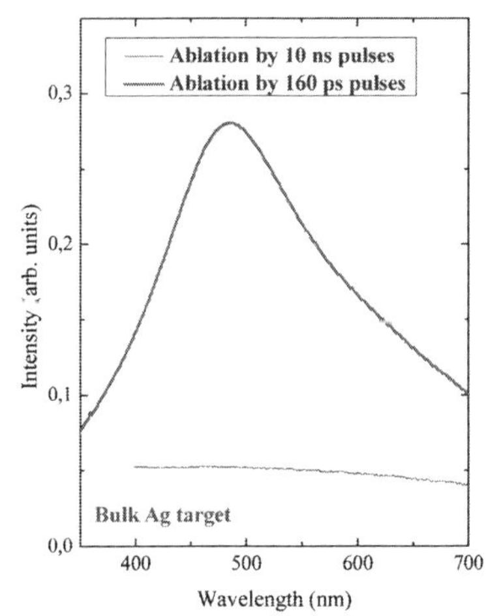

(b)

Figure 7.13 Absorption spectra of deposits generated from (a) silver nanoparticle-containing target and (b) bulk silver in the case of ablation by 10 ns (thin curves) and 160 ps (thick curves) heating pulses. Reproduced with permission from Ref. 55. Copyright 2013 American Physical Society.

amount of free electrons, which led to phase mismatch between the driving and harmonic waves.

These morphological studies regarding the nanoparticles in the plasma debris were supported by measurements of the absorption spectra of the deposits. For these purposes, we deposited the debris from plasma plumes on the glass substrates and then measured the transmittances of depositions. These measurements are illustrated in Fig. 6 for the case of silver. Figure 7.13a shows the appearance of strong absorption bands in the vicinity of 400–500 nm in the cases of the debris from the ablation of 25 nm silver nanoparticles using 160 ps and 10 ns pulses. These absorption bands are ultimately associated with the surface plasmon resonance (SPR) of this metal and its position and width are related with the shape, size, structure, and assembly of the nanoparticles [57].

One can note that the surface plasmon resonance appearing in the case of ablation using 160 ps pulses could be attributed to smaller clusters (3–5 nm) rather than 25 nm nanoparticles. Our attempts to reproduce this curve by dissolving original (unablated)

Ag nanoparticles in different liquids and measuring the absorption spectra of this suspension did not show the appearance of absorption band associated with the plasmon resonance of silver. In the meantime, SEM measurements did not reveal the presence of small nanoparticles in the debris due to insufficient resolution of this method, while TEM images demonstrate the presence of particles with the sizes less than 10 nm in the original powders (see Fig. 7.9a). Probably, the ablation of silver nanoparticles could cause some disintegration of the part of ablating particles and appearance of smaller nanoparticles. The absorption spectra from deposited debris of copper nanoparticles also showed the presence of surface plasmon resonance in the vicinity of 600 nm, though not so pronounced. No resonance absorption was observed in the case of the debris from ablating aluminum nanoparticles.

The spectral measurements of the depositions from bulk silver (corresponding to the conditions presented in Fig. 7.12a) also show the presence of strong absorption area in the range of 475 nm in the case of ablation using 160 ps pulses (Fig. 7.13b, thick curve), while the debris from 10 ns induced plasma did not show any spectral peculiarities (Fig. 7.13b, thin curve).

7.3.3 *Harmonic Generation from Nanoparticle-Containing Plasmas*

I have already mentioned the application of powdered targets for ablation-induced HHG by applying the C_{60} powder [2]. We extend the application of nanomaterials of considerably larger sizes for HHG in plasma plumes using 1 kHz laser. The use of relatively high-sized nanoparticles (of ~20 nm in diameter) can enhance the HHG yield due to higher concentration of neutrals compared with 0.7 nm fullerenes. Next, we have the opportunity to compare the HHG from monoparticles (atoms and ions) and agglomerates of atoms of the same element of periodic table. Finally, we can compare the nonlinear optical response from small (~20 nm) and large (~100–150 nm) nanoparticles. The application of high pulse repetition rate laser increases the average power of generating harmonics in these configurations and requires less time for collecting all necessary data on the nonlinear optical properties of these species.

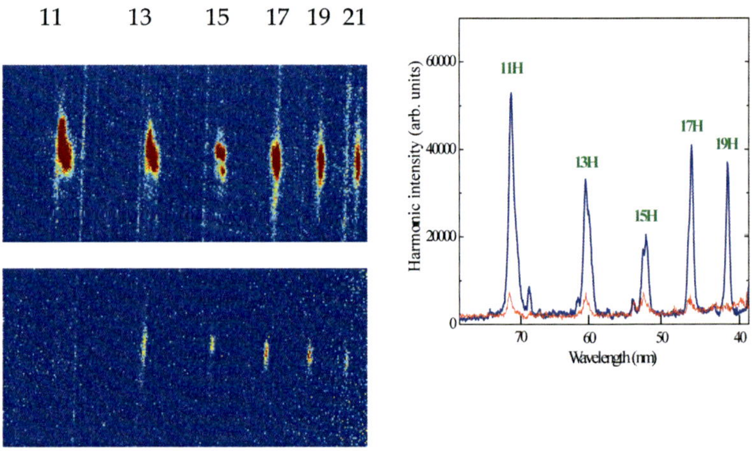

(a) (b)

Figure 7.14 (a) Raw images of harmonic spectra in a laser plasma produced by ablation with 160 ps pulses of 25 nm Ag nanoparticle based target (upper panel) and bulk Ag target (bottom panel). (b) Relative intensities of silver monoparticles (thin curve) and nanoparticles (thick curve) harmonics. Reproduced with permission from Ref. 55. Copyright 2013 American Physical Society.

The comparison of HHG yields from the plasmas produced on the bulk silver and 25 nm silver nanoparticles-containing target is presented in Fig. 7.14. The ablation conditions remained same for these two targets. The same can be said about other parameters of experiments (distance between the target surface and driving beam, pulse duration of heating radiation, delay between heating and driving pulses, etc.). Figure 7.14a shows the CCD images of harmonic spectra generating between 40 and 70 nm from nanoparticle- and monoparticle-containing Ag plasmas, while Fig. 7.14b presents the relative intensities of those harmonics. The ratio I_{nano}/I_{mono} between harmonics in this case was in the range of 5 to 7. It is also seen that the divergence of nanoparticle harmonics, as well as their spectral width, were considerably larger than those of monoparticle harmonics. The image of nanoparticles-induced harmonics also contains the spectral lines of ions generating during laser ablation of nanoparticles (Fig. 7.14a, upper panel). One can note that the appearance of ionic lines in these spectra is related with the presence of relatively large amount of free electrons in the plasma plume.

Their role in worsening of the conditions of HHG has been a topics of studies during long time both for gas harmonics and plasma harmonics. It has been reported [4] that strong harmonic emission from monoparticle plasmas was obtained when no plasma lines from excited ions appear in the XUV spectra. The simultaneous appearance of strong harmonics and spectral lines points out the growing influence of large clusters in creation of favorable conditions for HHG even in the presence of some amount of free electrons. The prevailing influence of the role of nanoparticles over free electrons in HHG, due to increased cross section of recombination of the accelerated electron and the local field enhancement in the presence of nanoparticles related with their plasmonic properties, is an interesting peculiarity of this process, which needs further investigation.

Even larger ratio between nanoparticle and monoparticle harmonics was observed in the case of Al clusters (of a mean diameter of 20 nm) and Al atoms and ions. The harmonics from monoparticles (Fig. 7.15a) had less divergence and spectral width at equal conditions of experiment, while nanoparticle harmonics (Fig. 7.15b) were much stronger. The ratio between harmonics in this case obtained at identical conditions was in the range of 7–11. The application of copper nanoparticles showed less efficient harmonics, which could be caused by less favorable conditions of plasma formation due to clusterization of large nanoparticles during laser ablation.

Notice that larger divergence of harmonics in the latter case could be attributed to the stronger influence of long trajectories of accelerating electrons in overall yield of harmonic radiation. This assumption could be related with the growth of cross section of recombination, when larger particles have higher probability to catch the returning electron moving along the long trajectory and recombine with further emission of harmonic photons. This peculiarity (i.e., larger divergence of harmonics) was observed for different nanoparticle-containing plumes. Notice that, in the case of aluminum nanoparticle plasma, we did not see the ionic lines

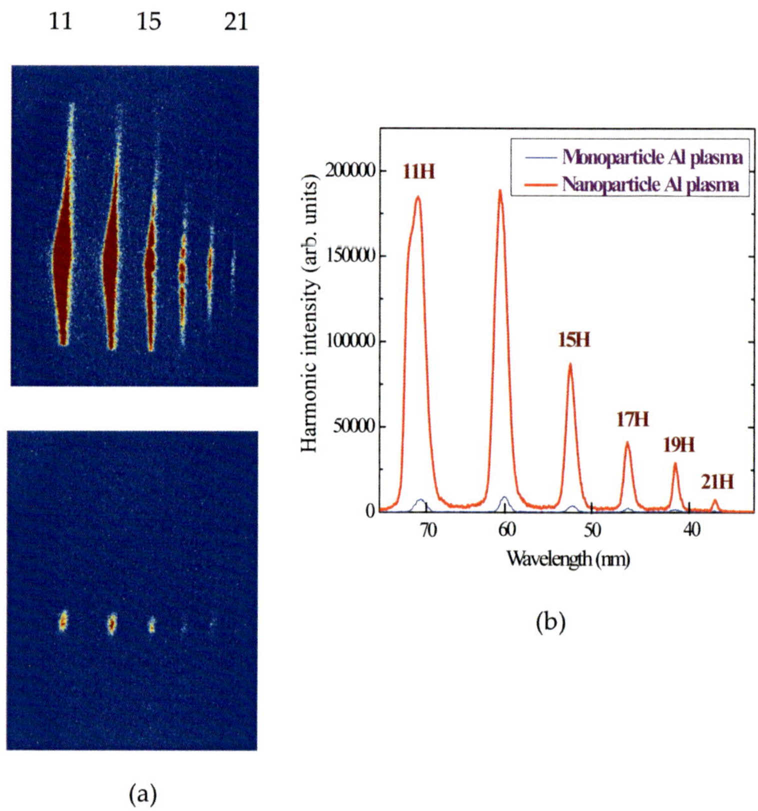

Figure 7.15 (a) Raw images of aluminum plasma harmonic spectra in the cases of ablation of 20 nm Al nanoparticles (upper panel) and bulk Al target (bottom panel). The conditions of experiments were the same for ablation by heating 160 ps pulses, focusing of femtosecond pulses, delays between pulses, and distances from the target surfaces. (b) Lineouts of the relative intensities of aluminum monoparticles (thin curve) and nanoparticles (thick curve) harmonics. Reproduced with permission from Ref. 55. Copyright 2013 American Physical Society.

alongside with nanoparticle harmonics, contrary to the case of Ag nanoparticle experiments. The ablation of superglue, without nanoparticles, did not show the harmonic generation and plasma emission at various experimental conditions, which confirms that the emitters of harmonics were associated with the nanoparticles of studied samples.

We observed only relatively low-order harmonics from nanoparticle-containing plasmas. This could be related with the origin of nanoparticles-induced HHG, which is assumed to occur from neutrals (i.e., neutral atoms of silver, aluminum, and copper nanoparticles). Notice that some previous studies of bulk silver harmonic spectra have shown the deep in the harmonic distribution around the 40–55 nm spectral range, while the shorter wavelength harmonics showed larger intensity. Nanoparticle harmonics did not showed same extension of harmonic cutoff as monoparticle harmonics. One can assume that harmonics in the former case were originated from neutrals of nanoparticles, while, in the latter case, the harmonics are generated from silver ions.

Our studies of the HHG in the plasmas containing 100–150 nm Ag nanoparticles and 100 nm Al nanoparticles showed some insignificant enhancement of harmonics compared with the case of plasma formation on the bulk aluminum and silver targets, though this enhancement was few times less compared with the case when we used smaller nanoparticles. Moreover, in some cases the intensities of harmonics in the former cases were weaker than those from monoparticle-containing plasmas. Probably, the application of large nanoparticles does not lead to the growth of harmonic yield due to decrease of local field effect and growth of the cross section of the absorption of generating harmonics inside the large clusters. The issue of the optimal sizes of the nanoparticles participating in harmonic generation still unresolved due to necessity in definition of the relative role of different competitive mechanisms (i.e., enhancing and diminishing processes) during HHG in clustered plasma.

This method of target fabrication could be very useful in the case of any powder-like species (fullerenes, metal nanoparticles, nanotubes, organic powder-like samples, etc.). Previously, even at 10 Hz pulse repetition rate, the ablation of static powdered nanotargets led to their rapid degradation and abrupt decrease of harmonic yield after few hundred shots [58]. In the case of the powder-like targets ablated by repetitive laser shots on the same spot at 1 kHz, the degradation of plasma plume is largely related with the unstable conditions of the ablating spot where the appearance of melted bath can considerably change the conditions of optimal plasma formation. After moving to a new spot, the previously irradiated target area cools down and becomes available (one rotation later) again for further ablation with approximately same harmonic yield. Meanwhile, one can note that the stability of harmonics from such powder-like targets is still considerably worse than from rotating bulk metal rods due to faster degradation of the target surface in the former case, while, in the latter case, the stability of harmonics during 10^6 shots has been achieved [5].

The comparative measurements of harmonics generated from three different plasmas containing Ag, Al, and Cu nanoparticles of approximately same sizes showed that Ag provides the highest conversion efficiency into the XUV. The silver superior performance could be attributed to the influence of the SPR of this metal. The wavelength of the second harmonic of the driving field (390 nm) almost coincides with the short-wavelength wing of the SPR band of silver (see Fig. 7.13). This coincidence can enhance the nonlinear susceptibility of the medium and favor the HHG yield. It is still unclear which nanoparticles play most important role in the enhancement of harmonic yield. From one hand, the increase of nanoparticle sizes leads to the growth of recombination cross section with returning electron. From another hand, generation of harmonics in the clusters containing huge amount of particles can be restricted by absorption of generating harmonic wave by neighboring atoms of cluster.

References

1. V. R. Bhardwaj, P. B. Corkum, D. M. Rayner, *Phys. Rev. Lett.* 93, 043001 (2004).
2. R. A. Ganeev, C. Hutchison, T. Witting, F. Frank, S. Weber, W. A. Okell, E. Fiordilino, D. Cricchio, F. Persico, A. Zaïr, J. W. G. Tisch, J. P. Marangos, *J. Opt. Soc. Am. B* 30, 7 (2013).
3. Y. Pertot, L. B. Elouga Bom, V. R. Bhardwaj, T. Ozaki, *Appl. Phys. Lett.* 98, 101104 (2011).
4. R. A. Ganeev, *J. Modern Opt.* 59, 409 (2012).
5. C. Hutchison, R. A. Ganeev, T. Witting, F. Frank, W. A. Okell, J. W. G. Tisch, J. P. Marangos, *Opt. Lett.* 37, 2064 (2012).

6. R. A. Ganeev, L. B. Elouga Bom, J. Abdul-Hadi, M. C. H. Wong, J. P. Brichta, V. R. Bhardwaj, T. Ozaki, *Phys. Rev. Lett.* 102, 013903 (2009).
7. J. Roth, F. Géahler, H.-R. Trebin, *J. Modern Phys. C* 11, 317 (2000).
8. P. B. Corkum, *Phys. Rev. Lett.* 71, 1994 (1993).
9. E. Fiordilino, V. Miceli, *J. Modern Opt.* 41, 1415 (1994).
10. J. Zhou, J. Peatross, M. M. Murnane, H. C. Kapteyn, I. P. Christov, *Phys. Rev. Lett.* 76, 752 (1996).
11. M. Lein, *Phys. Rev. A* 72, 053816 (2005).
12. M. A. L. Marques, C. A. Ullrich, F. Nogueira, A. Rubio, K. Burke, E. K. U. Gross, *Time-dependent Density Functional Theory,* Springer, Berlin (2006).
13. K.Yabana, T. Otobe, J. I. Iwata, in *Progress in Ultrafast Laser Science I,* K. Yamanouchi, S. L.Chin, P. Agostini, G.Ferrantre, eds., 77, Springer, (2006).
14. D. Bauer, F. Ceccherini, A. Macchi, F. Cornolti, *Phys. Rev. A* 64, 063203 (2001).
15. M. J. Pushka, R. M. Nieminen, *Phys. Rev. A* 47, 1181 (1993).
16. J. L. Martins, N. Troullier, J. H. Weaver, *Chem. Phys. Lett.* 180, 457 (1991).
17. D. Cricchio, P. P. Corso, E. Fiordilino, G. Orlando, F. Persico, *J. Phys. B* 42, 085404 (2009).
18. P. P. Corso, E. Fiordilino, G. Orlando, F. Persico, *Recent Res. Develop. Phys.* 5, 1445 (2004).
19. A. D. Bandrauk, S. Chelkowski, D. J. Diestler, J. Manz, K.-J. Yuan, *Phys. Rev. A* 79, 023403 (2009).
20. M. H. Beck, A. Jäckle, G. A. Worth, H.-D. Meyer, *Phys. Rep.* 324, 1 (2000).
21. H.-D. Meyer, U. Manthe, L. S. Cederbaum, *Chem. Phys. Lett.* 165, 73 (1990).
22. U. Manthe, H.-D. Meyer, L. S. Cederbaum, *J. Chem. Phys.* 97, 3199 (1992).
23. R. A.Ganeev, G. S. Boltaev, R. I. Tugushev, T. Usmanov, M. Baba, H. Kuroda, *Eur. Phys. J. D* 64, 109 (2011).
24. R. A.Ganeev, H. Singhal, P. A. Naik, J. A. Chakera, A. K. Srivastava, T. S. Dhami, M. P. Joshi, P. D. Gupta, *Appl. Phys. B* 100, 581 (2010).
25. R. A. Ganeev, L. B. Elouga Bom, M. C. H. Wong, J.-P. Brichta, V. R. Bhardwaj, P. V. Redkin, T . Ozaki, *Phys. Rev. A* 80, 043808 (2009).
26. S. Kim, I.-Y. Park, J. Choi, S.-W. Kim, in *Progress in Ultrafast Laser Science VI* . K. K. Yamanouchi, G. Gerber, A. D. Bandrauk, eds., 147, Springer, Berlin (2006).
27. McPherson, G. Ginson, H. Jara, N. Johann, I. A. McIntyre, K. Boyer, C. K. Rhodes, *J. Opt. Soc. Am. B* 4, 595 (1987).

28. M. Ferray, A. L'Huillier, X. F. Li, L. A. Lompré, G. Mainfray, G. Manus, *J. Phys. B: At. Mol. Opt. Phys.* 21, L31 (1988).
29. L. A. Lompré, A. L'Huillier, M. Ferray, P. Monot, G. Mainfray, G. Manus, *J. Opt. Soc. Am. B* 7, 754 (1990).
30. Y. Akiyama, K. Midorikawa, Y. Matsunawa, Y. Nagata, M. Obara, H. Tashiro, K. Toyoda, *Phys. Rev. Lett.* 69, 2176 (1992).
31. P. A. Franken, A. E. Hill, C. W. Peters, G. Weinreich, *Phys. Rev. Lett.* 7, 118 (1961).
32. K. Miyazaki, H. Takada, *Phys. Rev. A* 52, 3007 (1995).
33. I. P. Christov, M. M. Murnane, H. C. Kapteyn, *Phys. Rev. Lett.* 78, 1251 (1997).
34. G. Tempea, M. Geissler, M. Schnürer, T. Brabec, *Phys. Rev. Lett.* 84, 4329 (2000).
35. A. Pirri, C. Corsi, M. Bellini, *Phys. Rev. A* 78, 011801 (2008).
36. J.-F. Hergott, M. Kovacev, H. Merdji, C. Hubert, Y. Mairesse, E. Jean, P. Breger, P. Agostini, B. Carré, P. Salières, *Phys. Rev. A* 66, 021801 (2002).
37. J. F. Reintjes, *Nonlinear Optical Parametric Processes in Liquids and Gases*, Academic Press, Orlando (1984).
38. R. A. Ganeev, M. Baba, M. Suzuki, H. Kuroda, *Phys. Lett. A* 339, 103 (2005).
39. R. A. Ganeev, M. Suzuki, T. Ozaki, M. Baba, H. Kuroda, *Opt. Lett.* 31, 1699 (2006).
40. T. D. Donnelly, T. Ditmire, K. Neuman, et al., *Phys. Rev. Lett.* 76, 2472 (1996).
41. J. W. G. Tisch, T. Ditmire, D. J. Fraser, et al., *J. Phys. B: At. Mol. Opt. Phys.* 30, L709 (1997).
42. S. X. Hu, Z. Z. Xu, *Appl. Phys. Lett.* 71, 2605 (1997).
43. J. W. G Tisch, *Phys. Rev. A* 62, 041802 (R) (2000).
44. C. Vozzi, M. Nisoli, J.-P. Caumes, et al., *Appl. Phys. Lett.* 86, 111121 (2005).
45. C.-H. Pai, C. C. Kuo, M.-W. Lin, et al., *Opt. Lett.* 31, 984 (2006).
46. C. Hubert, L. Billot, P.-M. Adam, R. Bachelot, P. Royer, J. Grand, D. Gindre, K. D. Dorkenoo, A. Fort, *Appl. Phys. Lett.* 90, 181105 (2007).
47. S. Kim, J. Jin, Y.-J. Kim, I.-Y. Park, Y. Kim, S.-W. Kim, *Nature*, 453, 757 (2008).
48. T. Hanke, G. Krauss, D. Traütlein, B. Wild, R. Bratschitsch, A. Leitenstorfer, *Phys. Rev. Lett.* 103, 257404 (2009).
49. K. D. Ko, A. Kumar, K. H. Fung, R. Ambekar, G. L. Liu, N. X. Fang, K. C. Toussaint, *Nano Lett.* 11, 61 (2011).

50. R. A. Ganeev, L. B. Elouga Bom, J. Abdul-Hadi, M. C. H. Wong, J. P. Brichta, V. R. Bhardwaj, T. Ozaki, *Phys. Rev. Lett.* 102, 013903 (2009).
51. R. A. Ganeev, M. Suzuki, M. Baba, M. Ichihara, H. Kuroda, *J. Appl. Phys.* 103, 063102 (2008).
52. R. A. Ganeev, L. B. Elouga Bom, T. Ozaki, *J. Phys. B: At. Mol. Opt. Phys.* 42, 055402 (2009).
53. H. Singhal, R. A. Ganeev, P. A. Naik, A. K. Srivastava, A. Singh, R. Chari, R. A. Khan, J. A. Chakera, P. D. Gupta, *J. Phys. B: At. Mol. Opt. Phys.* 43, 025603 (2010).
54. R. A. Ganeev, *Laser Phys.* 18, 1009 (2008).
55. R. A. Ganeev, C. Hutchison, M. Castillejo, I. Lopez-Quintas, F. McGrath, D. Y. Lei, J. P. Marangos, *Phys. Rev. A* 88, 033803 (2013).
56. M. López-Arias, M. Oujja, M. Sanz, R. A. Ganeev, G. S. Bolataev, N. K. Satlikov, R. I. Tugushev, T. Usmanov, M. Castillejo, *J. Appl. Phys.* 111, 043111 (2012).
57. V. Amendola, O. M. Bakr, F. Stellacci, *Plasmonics* 5, 85 (2010).
58. R. A. Ganeev, H. Singhal, P. A. Naik, J. A. Chakera, A. K. Srivastava, T. S. Dhami, M. P. Joshi, P. D. Gupta, *J. Appl. Phys.* 106, 103103 (2009).

Summary

Perspectives of Plasma Harmonics

The quest for new plasma media that would favor the enhancement of an individual harmonic allows further enhancement of harmonic conversion efficiency. The production of a single high-intensity harmonic (rather than a group of harmonics of equal intensity in the plateau region) would open up the way to the practical application of these coherent short-wavelength radiation sources. Because resonantly enhanced harmonics have already been observed in several plasma media, there are strong grounds to believe that similar conditions will be discovered for other plasma formations. The generated harmonic wavelength may then be tuned to the transitions with high oscillator strength by wavelength tuning of the driving laser system, as well as by varying the chirp of the laser radiation. Application of ablated nanoparticles and clusters for HHG can also enhance the yield of harmonics in the XUV range. Further improvements in HHG conversion efficiency and harmonic extension require a systematic study of the influence of various plasma and laser parameters on ablation harmonics. Many new peculiarities of plasma harmonics emerged during last few years allow expecting further extension of our knowledge of materials properties using this powerful tool of nonlinear spectroscopy.

Although the first stage of these successful studies was entirely focused on the improvements of harmonic yield from plasma, at current stage of knowledge of the high-order nonlinear optical processes in ablation plume, one can consider this method as a new tool for materials science. Thus, the search of the dual role of plasma HHG as a method for efficient coherent XUV light generation and of materials probing is a milestone of proposed project. The project is aimed at the enhancement of HHG efficiency from laser ablation produced on the surfaces of solid-state materials and applications of plasma harmonic studies for analysis of spectral and structural properties of materials.

Future studies are aimed on an analysis of the HHG spectroscopy conducted using a low-excited laser-produced plasma, which has already demonstrated several new approaches to the problem of increasing the HHG efficiency in comparison with the efficiency of conventional HHG in gases. At the same time, it is obvious that gas HHG exhibits much higher harmonic orders (in comparison with the plasma HHG obtained to date). Without making excessively optimistic predictions that the orders of in-plasma-generated harmonics will exceed the in-gas harmonic orders in the near future, we will focus primarily on the realization of new techniques for increasing the wavelength-converted radiation intensity with the use of plasma plumes, which may hardly be realized in conventional gas HHG. At the same time, one can explore various optical, nonlinear optical, and structural properties of materials using the laser ablation induced high-order harmonic generation spectroscopy (LAIHOHGS).

One can expect an improvement of plasma HHG due to the double excitation of laser-produced plasmas, optimization of the longitudinal harmonic generation scheme in the laser plume, optimization of nanostructured plasmas, use of multicomponent plasma plumes, formation of quasi-phase matching conditions for the waves in complex extended plasmas, provision of a regime of waveguide pump propagation through the plasma medium, feasibility analysis of attosecond pulse generation in laser-produced plasma, use of mid-IR pulses for harmonic extension, stabilization of plasmas and harmonics characteristics using rotating and moving targets, search of the attosecond pulse generation by achieving the continuum in the harmonic emission near the cutoff, which allows for the first time the generation of ultrashort laser pulses in the plasma plumes, and so on. All these concepts are new and did not apply previously for the improvement of plasma harmonic generation efficiency. The joint implementation of new and old (application of the clusters with controllable and variable sizes, resonance-induced harmonic enhancement, two-color pump-induced enhancement of even harmonics, search of the influence of multi-electron dynamics of complex clusters, such as fullerenes and nanotubes, on the plasmon resonance-induced growth of few harmonics in the XUV range, and so on) methods and approaches in the case of plasma media allows making the advance to the state-of-the-art within the project field, which will lead to the establishment of a new method of material science, laser ablation induced high-order harmonic generation spectroscopy. It follows from the above that investigations in this area of nonlinear optics are making rapid strides and may bring new success in the near future.

It is worth noting that, currently, nonlinear spectroscopy involving gas HHG is trying to deal with problems such as (a) extension of the HHG spectroscopy based attosecond structural imaging technique to image nuclear re-arrangements induced by localized hole excitations, (b) application of strong-field ionization to create localized hole excitations and study their attosecond dynamics in polyatomic molecules, (c) search of selective imaging of hole dynamics induced by the removal of, for example, inner-valence electrons using the XUV-initiated HHG technique, and (d) development of multi-dimensional HHG spectroscopy capable of following energy flow between different molecular modes over multiple femtosecond time-scale. The proposed LAIHOHGS allows adding some important impulse to those studies, by using the peculiarities of plasma HHG.

In the nearest future, the materials studies technique using the LAIHOHGS, which exploits the spectral and structural properties of various ablated solid-state materials using the propagation of short laser pulse through laser-produced plasma and generation of high-order harmonics, will be developed. The main goals of these studies will be a search of new schemes of high-order harmonic generation using the laser ablation, generation of femtosecond and attosecond XUV pulses through the HHG in plasma, enhancement of harmonic efficiency in these schemes, and the materials studies using the LAIHOHGS, which exploits the spectral and structural properties of various solid-state materials through their ablation and further propagation of short laser pulse through laser-produced plasma and generation of high-order harmonics. The important key step is putting, through ablation, solid molecules in gas phase at densities sufficient for HHG spectroscopy measurements. More concrete tasks and methodologies for achieving these goals are as follows:

New approaches in harmonic generation from the laser plumes: The researchers will carry out the analysis of the harmonic generation of ultrashort (multi- and few-cycle) laser pulses from the plasma produced on the surface of various targets at different conditions of experiment (driving laser wavelength and pulse duration, pulse repetition rate, excitation of plasma by nano-, pico-, and femtosecond pulses, variable plasma parameters, single atom, molecular, and cluster-containing targets, and so on). Various new ideas for further amendments of this process will be examined. Among them are harmonic generation in plasma using two-color pump configuration in the case of commensurate and noncommensurate wavelength sources in the mid-infrared and ultraviolet ranges; search for the conditions of attosecond pulse trains generation using various gating techniques; measurements of pulse duration of the plasma harmonics generated by few-cycle pulses; studies of HHG from various clusters appearing in situ during laser ablation of targets; manipulation of plasma parameters by the second heating pulse propagating through the plasma simultaneously with driving femtosecond pulse; analysis of phase-matching conditions for plasma HHG. The influence of molecular orientation on the harmonic output from molecules-containing plumes will be studied.

Analysis of plasma characteristics at "optimal" and "non-optimal" conditions of HHG: Studies of plasma composition using scanning electron microscopy; time-of-flight mass spectrometry of plasma components at the conditions of efficient HHG; development of stable plasma formation technique at high pulse repetition rate and analysis of HHG at these conditions.

Development of ablation-induced HHG spectroscopy of various solid materials after their evaporation through laser ablation, particularly, the orientation-induced response of large ablated molecules and nanoparticles, and the time-resolved pump-probe analysis of complex plasmas containing various molecular structures.

Search of new schemes for resonantly enhanced harmonics. Further studies of resonance enhancement, increase of the HHG efficiency, and extension of the wavelength range for such enhancement. Use of two-color and mid-IR driving pulses for achieving the resonance conditions with the ionic transitions of various plasmas in different spectral ranges.

Development of various new approaches in plasma HHG, including the quasi-phase matching in extended plasma plumes, studies of the dynamics of aggregation and disintegration of clusters through their nonlinear optical response, comparative analysis of gas and plasma HHG at different conditions, joint application of gas and plasma HHG for the studies of gaseous and ablated species, creation of the conditions for single harmonic generation in the XUV range using few-cycle pulses, generation of continuum in the plasma harmonic spectra in different ranges of XUV using single and double gating techniques, and so on.

The extension of highest cutoff energy of generated harmonics using the interaction of intense laser radiation with doubly charged ions. Analysis of harmonics extension in the case of longer wavelength (1200–2000 nm) driving radiation.

Harmonic generation from laser plumes containing nanoparticles. The influence of the surface plasmon resonances of clusters in plasma on the resonance-induced enhancement of harmonics. The analysis of plasmonic properties of carbon nanoparticles for enhancement of harmonics. The studies of the indirect involvement of clusters in HHG, when they did not participate as the harmonic emitters, but rather enhance the local field, analogous to recently reported experiments and calculations using the gold nanostructures enhancing gas HHG.

Application of mid-infrared radiation to study the dynamics of nonlinear optical response of ablated solids compared with commonly used 800 nm class lasers for plasma HHG, including the studies of extended harmonics at comparable conversion efficiency with shorter wavelength laser sources, and a search of new opportunities in improvement of HHG conversion efficiency in the mid-IR range. The nonlinear optical interaction of ablated molecules with intense 800 nm and longer wavelength laser sources. In the latter case (min-IR pulses), this interaction up-converts the frequency of the incident radiation by a factor of nearly 10^2 yielding plasma harmonics of the incident radiation extending to the spectral region below 10 nm. The harmonic emission results from the radiative recombination of an electron, liberated by ionization of a molecule and accelerated by the IR field, with the hole left in the ablated molecule. Spectra, phases, and polarization of the emitted radiation encode the shape, location, and momentum of the hole during recombination as well as ultrafast changes of the molecular structure that have occurred in the time between ionization and recombination.

A separate topic of these studies will be the application of ablation plumes for attosecond science. Measurement of physical processes with a temporal resolution approaching 10^{-16}s has emerged in the last few years as one of the most exciting frontiers in physical science. Such measurements make accessible the ultrafast dynamics of correlated electronic motion that underpin the first moments of a wide range of physical and chemical processes, for example, in photochemical reactions, radiation damage in biomolecules, and in converting light energy into chemical energy. This science demands the most advanced technology and in particular suitable light sources of exceptionally high bandwidth (>10 eV) to support the ultra-high temporal resolution. Currently, the primary technique to do this is high-order harmonic generation in gases. The achievements in this field include generation of isolated attosecond pulses from atomic gases and use of the time-energy encoding of the HHG spectrum to deduce sub-femtosecond nuclear and electronic dynamics in the molecule. Nevertheless, the capabilities for attosecond measurements are severely limited by several factors, including (a) low photon yield in the generation of isolated attosecond pulses, and (b) the limited range of molecules that can be obtained in gas phase at densities sufficient for HHG.

The purpose of this research is to enable various groups to fully develop world-leading capabilities in attosecond sciences using these newly identified plasma HHG spectroscopy methods. Although the work is challenging, I believe that the present expertise in this field has the possibility to make a step-change advance in plasma harmonics induced attosecond science. The main objectives of this research are:

To test whether, as expected from calculation, the resonance-enhanced HHG in Mn, In, and Zn ablation plumes leads to isolated sub-femtosecond pulses using the attosecond streaking technique.

To utilize the isolated sub-femtosecond pulses from these metals in pump-probe measurements at surfaces and in molecules.

To develop optimal conditions for ablation plumes of high density for HHG studies of intact ribonucleic acid and deoxyribonucleic acid bases (e.g., uracil and thymine) to allow the first steps for attosecond investigation of electron's dynamics in these molecules.

Currently, a number of scientific groups (in United Kingdom, Ireland, Japan, France, Korea, United States, India, Germany, Spain, and Italy) became interested in the development of laser ablation induced plasma sources as the media for harmonic generation and as a tool for LAIHOHGS. A successful implementation of new ideas in this field requires the appropriate skill, which has been grown during recent years.

All the above-proposed goals and tasks are new and did not explore previously, excluding few approaches, which are currently under consideration in various laboratories worldwide. I believe that the novelty of the majority of proposed ideas and the availability of resources allow expecting a successful realization of above tasks, which further extend our knowledge in the field of materials science using the nonlinear spectroscopy tools.

Index